Detlef Gysau

Lab Automation and Digitalisation in Coatings

Cover image: Summit Art Creations – stock.adobe.com

Bibliographische Information der Deutschen Bibliothek
Die Deutsche Bibliothek verzeichnet dies Publikation in der Deutschen Nationalbibliografie; detaillierte bibliografishe Daten sind im Internet über http://dnb.d-nb.de abrufbar.

Detlef Gysau
Lab Automation and Digitalisation in Coatings
Hanover: Vincentz Network, 2025
European Coatings Library
ISBN 3-7486-0799-7
ISBN 978-3-7486-0799-1

Please ask for our book catologue:
Vincentz Network, Plathnerstr. 4c, 30175 Hanover, Germany
Tel. +49 511 9910-033, Fax +49 511 9910-029
E-mail: books@european-coatings.com, www.european-coatings.com

Layout: Vincentz Network, Plathnerstr. 4c, 30175 Hanover, Germany
Printing: Gutenberg Beuys, Hanover, Germany

European Coatings Library

Detlef Gysau

Lab Automation and Digitalisation in Coatings

1 Preface

The world of laboratory automation is at a turning point: advances in technologies such as robotics, artificial intelligence and data integration have revolutionized the ways in which research and development are conducted. At the same time, however, I observe that many companies and professionals lack a clear direction on how to effectively and sustainably integrate these innovations into their labs.

I was driven by the desire to make this knowledge more accessible and to build a bridge between technological progress and practical implementation. With my many years of experience in laboratory process automation, combined with my background in various industries, I have been able to identify numerous challenges and best practices. I still remember my first exposure to a high throughput system of Labman at AkzoNobel in Slough, UK, in 2008. That's when I caught the virus and started implementing it at Omya.

My aim with this book is to share knowledge. To make the complex possibilities and challenges of laboratory automation understandable for a broad audience. I also want to provide inspiration and motivate readers to see automation not just as a tool, but as a strategic advantage. The presentation of practice-oriented solutions ranges from planning to implementation, with a special focus on efficiency, quality and sustainability. This book is aimed at anyone who is professionally involved in any way with laboratory processes for coating materials. Beginners and students will get a comprehensive overview of the field, while experienced developers will find practical details relevant for solving their daily challenges in the future.

This book is intended to help guide companies and laboratories on their way into the future. As I firmly believe that laboratory automation is not only a tool for increasing efficiency, but also a pioneer for innovation and scientific progress. For this reason, I have used a lot of visual material to describe the possibilities that already exist today in order to show the reader in a simple way that the introduction of automated laboratory processes also requires a rethink. Manual processes and practices used up to now cannot always be automated 1:1. This is particularly the case for antiquated, historical measuring and testing methods, which are being replaced by modern instrument technology for characterizing the properties of paints and coatings.

It is important to mention that in my more than 15 years of experience in laboratory automation, there was never any thought of cutting jobs. Unfortunately, this is often the first thought in many people's minds. In reality, however, many examples from different industries and professions do not show job cuts but rather changes in activities. The focus of automation is primarily on shortening time to market, standardization, reproducibility and the elimination of mindless routine work. In recent years, the baby boomer effect has also played a role. It is becoming increasingly difficult to replace this retiring generation with new, young skilled workers. This is also spurring the introduction of automation in the laboratory, as well as in many other company processes. This change will continue in the future. The frontrunners in laboratory automation are already starting to use artificial intelligence and machine learning due to the large amounts of data generated.

This book can be seen as a step towards self-employment. Through my company, PERFECO Consulting Gysau, I focus on performance, economy, and ecology, offering vendor-neutral consulting and expertise to the industry.

Detlef Gysau
Full/Switzerland, March 2025

Acknowledgement

What motivated me to write a specialist book on laboratory automation? I admit it, I was motivated by various feminine sources. The request to write the book came from Vincentz Network, who has already supported me with my filler books, now in their 3rd edition. My wonderful wife Jacqueline also played a key role in my decision. She has always had my back in my professional career and encouraged me in all my activities. At the same time, she has brilliantly managed our family with both children and many animals, from dogs to cats, goats, chickens and horses. Dear Jacqueline – you deserve my deepest thanks! I would also like to gratefully thank my fabulous children, Gian-Flurin and Mica-Ladina, for their continued understanding and patience when I was not always available (similar to writing my other books).

I would like to express my deep thanks to Hendrik Hustert, Hermine Riegler, Ian Riley, John Hesford, Natalie Kainz, Roland Emmerich, Tobias Burk and Ulf Stalmach in alphabetical order, as well as the companies Anton Paar, Chemspeed Technologies, Füll Lab Automation, Labman Automation and Orontec for their uncomplicated support with technical information and images. My thanks also go to all my companions who have supported and encouraged me in any way in my professional career.

Contents

1 Introduction

1.1 History of automation

The first making of paints dates more than 100,000 years back and was discovered in South Africa approximately one decade ago [1]. The paints were exemplary dedicated for cultural purposes such as paintings in caves. Life was very different to what it is nowadays and there was no need to rush either for making or applying the paint. The industrial revolution was still far away as well as industry 4.0 and lab 4.0.

Times have changed, not only but tremendously with the beginning of the industrial revolution in the 18th century. Industry and society strove for growth by the revolutionary technology enabling mass production. And thus, the need for higher speed was also necessary for development and innovation in the field of science. Tools such as stirrers became operated by transmission received from water or steam power. Later, these became useable and thus mobile at other locations by electrification. It is hard to say that this is automation with the understanding of the 3rd millennium after Christi.

Before electronic components became widely available after World War II, laboratory automation was constructed by end users and designed for specific tasks such as filtration and washing operations. The earliest mention of automation in the United States chemical literature dates to 1875, when a device for unattended washing of filtrates was announced, see Figure 1.1.

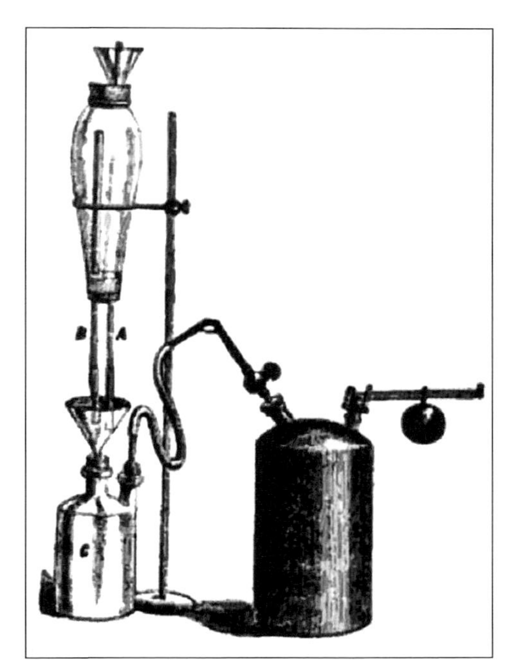

Figure 1.1: The filter washer described by Stewens in 1875 used a lamp chimney with water and closed at both ends to drip rinse water over a filtrate. The canister at the right of the device is a steam trap that was used to create vacuum
Source: Google Books

In the following years, a small number of automated devices were sold, including large mills for preparing coal samples. By 1900, power plants began using automated carbon dioxide analysis, see Figure 1.2 and Figure 1.3.

Around the time of World War I, the development of electrical equipment for conductivity measurements made possible the first commercial automated gas analyzers for laboratory and field use. The growth of industrial production in the 1920s led to a need for automated test equipment, and the growing rubber industry was among the most successful early adopters. The beginning of the Second World War saw a further surge in the development of automation solutions in process control. The background to this was, on the one hand, the increased requirements to produce war-related goods and, on the other hand, the shortage of qualified workers [2]. Particular attention was paid to the development of semi-automatic distillation apparatus [3]. By the end of World War II, the use of automated systems had become routine in the chemical industry. Electronic components were increasingly used to control valves. The development of automated titrators was advanced; in 1948, a device was developed that used a motor-driven syringe to add the titrant. Innovative technologies in liquid dosing were essential for the further development of laboratory automation. In 1957, Schnitger developed a new type of pipette that already had all the features of today's modern piston-stroke pipettes. Today's mechanically adjustable micropipettes can be traced back to a development by *Gilson*, which he patented in 1974 [4]. Technical advances in the development of small motors and valves led to the introduction of semi-automatic syringe-based pipetting systems in the 1970s. The development of microprocessor technology enabled the creation of program sequences to control the motors and valves, leading to the first fully automated pipetting systems. In the 1980s, further developments in electrical engineering resulted in the first automated liquid handling systems.

In the early 1980s a revolution took place; the first fully automated laboratory was opened by *Dr. Masahide Sasaki* [5, 6]. One of the world's first clinical automated laboratory management systems were developed at the University of Nebraska Medial Center by *Dr. Rod Markin* in 1993 [7]. In addition to the requirements of clinical laboratories, the development of high-throughput screening (HTS) methods in the pharmaceutical industry has been of particular importance for the development of laboratory automation. Parallel sample processing has been increasingly used in bio screening automation. Test methods based on microtiter plates were first introduced in 1986 [8]. With interchangeable hands, the systems could perform various laboratory processes such as pipetting, washing the plates, or adding reagents. The use of articulated robots was a very costly way of automating such processes and therefore not generally applicable. Numerous companies have therefore developed specialized liquid handling systems based on a "Cartesian" robot structure. The first 96-channel pipetting system was developed by TomTec in 1990 [9], followed later by a variant with a 384-pipetting head. Developments in the field of collaborative robots in recent years are increasingly enabling the use of cost-effective automation solutions in small and medium-sized enterprises, see Figure 1.4.

Figure 1.2: The "Sturtevant Automatic Coal Crusher" could reduce large pieces of coal to a size more suited for laboratory analysis and partition out a representative sample. This greatly improved the efficiency of coal analysis
Source: Google Books

The automation of paint labs, like many other industries, began in the mid-20th century. The specific timeline for the introduction of automation in paint labs can vary depending on the level of automation and the processes involved. In the early stages, automation may have been limited to simple machinery and equipment to aid in mixing and dispensing paint ingredients. Parallel operation of stirrers in a lab refers to the practice of using multiple stirrers simultaneously to mix or agitate multiple samples or solutions at the same time. This technique is commonly employed in laboratories, especially when researchers or technicians need to process multiple samples efficiently and uniformly. The parallel operation of stirrers offers several benefits:

- **Timesaving:** Using multiple stirrers allows researchers to process several samples simultaneously, reducing the overall time required for experiments or preparations.
- **Consistency:** Parallel operation ensures that all samples experience similar stirring conditions, resulting in more consistent and reproducible experimental outcomes.
- **Increased throughput:** Laboratories with a high sample workload can benefit from parallel operation as it enables the processing of a larger number of samples in a shorter time frame.
- **Resource optimization:** Instead of dedicating one stirrer for each sample, parallel operation enables researchers to use fewer stirrers while achieving the same results.
- **Flexibility:** Different stirrers can be set to different speeds and configurations, allowing researchers to tailor the stirring conditions to the specific requirements of each sample.

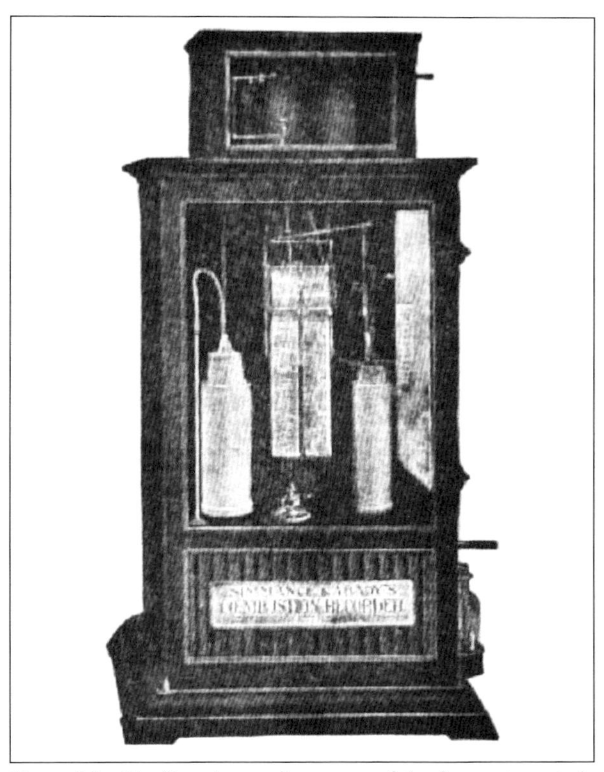

Figure 1.3: The "Autolysator" was one of the first automated analysis instruments ever sold commercially. This 1912 model used a complex system of clockwork, counterweights, and a chart recorder to provide real-time measurements of carbon dioxide in flue gas Source: Google Books

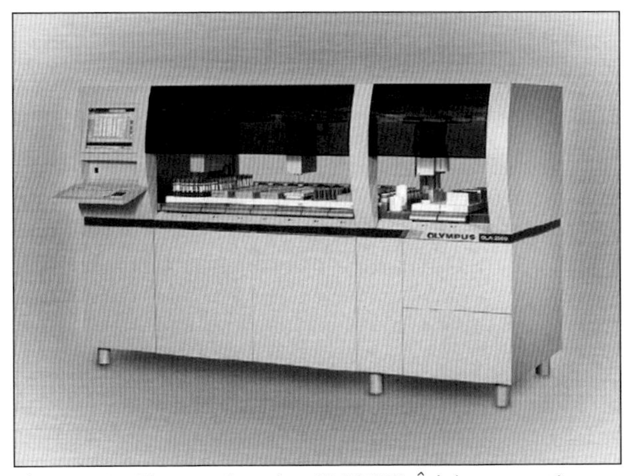

Figure 1.4: Olympus America OLA2500 Ô lab automation system has a maximum sorting throughput of 800 tubes per hour and aliquoting throughput of 650 tubes per hour. Source: Olympus America, INC., Center Valley, PA

When implementing parallel operation of stirrers, it is essential to ensure that each stirrer is properly calibrated, and the stirring conditions (such as speed, direction, and time) are standardized for all samples. Additionally, proper lab safety protocols should be followed to prevent any accidents or mishaps during the simultaneous operation of multiple stirrers.

As technology advanced, more sophisticated automated systems were developed to handle various aspects of paint manufacturing, quality control, and testing. In the early 21st century, with the rapid advancements in robotics, computer control systems, and artificial intelligence, paint labs saw a significant increase in automation capabilities. These advancements allowed for greater precision in color matching, faster production rates, improved quality control, and reduced human intervention in certain tasks. It's worth noting that the adoption of automation in paint labs, as in other industries, is an ongoing process, and new technologies continue to be integrated to optimize manufacturing processes and overall efficiency. In the meantime, automation in paint labs is well-established with large, financially strong paints and coatings manufacturers and the corresponding raw material industry, and it's likely that further advancements have occurred since then.

1.2 Needs for laboratory automation

Laboratory automation refers to the use of technology, robotics, and software to perform various tasks and processes in a laboratory setting. There are several reasons why laboratory automation is increasingly adopted. Some of the drivers, without representing completeness, for the increasing interest and thus the next step of the digital transformation are highlighted in the following sub chapters.

1.2.1 Increased efficiency

Lab automation allows for the simultaneous processing of multiple samples or experiments, significantly increasing the number of tests or analyses that can be conducted in a given timeframe. Automation can significantly speed up repetitive and time-consuming tasks, allowing scientists and researchers to focus on more complex and creative aspects of their work. Automated systems can work around the clock without breaks, leading to higher productivity.

1.2.2 Improved accuracy and precision

Improved accuracy and precision are crucial goals in scientific research, testing, and development. Accuracy refers to the closeness of a measurement or result to the true value, while precision refers to the consistency and repeatability of measurements. Automated equipment can carry out tasks with high precision and accuracy, leading to more reliable and reproducible results. Automation often includes integrated data collection and analysis, leading to more accurate and reliable data. This can improve the quality of research and decision-making processes.

1.2.3 Enhanced data quality

Enhanced data quality refers to the improvement and optimization of various attributes and characteristics of data to ensure that it is accurate, reliable, consistent, and fit for its intended purpose. Data quality is a critical aspect in scientific research and laboratory environments. When data quality is enhanced, it leads to greater confidence in the data's accuracy and the insights derived from it. Automation reduces the potential for manual data entry errors and ensures that data is recorded consistently and accurately. This leads to better data quality and more robust scientific conclusions. Thus, it is crucial for making informed decisions, drawing meaningful insights, and deriving reliable conclusions from data-driven analyses. It requires careful attention to data collection methods, processes, validation, and maintenance throughout the data lifecycle.

1.2.4 High throughput

High throughput in a laboratory context refers to the ability to process many samples, experiments, or analyses simultaneously in a relatively short period of time. It involves automating and streamlining laboratory processes to achieve a rapid and efficient workflow. High throughput is particularly valuable in research, diagnostics, drug discovery, and other industries where the volume of samples or data to be processed is substantial. The goal of high throughput is to increase efficiency, reduce turnaround times, and accelerate the pace of scientific discovery, development, and decision-making. While high throughput offers numerous benefits in terms of speed and efficiency, it is important to carefully design and validate high-throughput processes to ensure data quality, reproducibility, and accuracy. Real high-throughput workflows require typically investment in specialized equipment, automation technology, and data management solutions.

1.2.5 Standardization

Standardization in laboratories refers to the establishment and implementation of consistent protocols and standard operating procedures (SOP), practices, and guidelines to ensure uniformity, reliability, and accuracy in experimental processes, data collection, and analysis. It involves defining clear and agreed-upon methods that all personnel follow when conducting experiments, handling samples, using equipment, and recording data. Standardization is essential to maintain the quality and integrity of research, ensure reproducibility of results, and facilitate comparisons between different experiments or laboratories. Standardization is particularly important in the paints and coatings industry, where accurate and reliable results are critical for making informed decisions that impact performance, durability, application but also health, safety, and regulatory compliance. It ensures that experiments can be repeated, validated, and built upon, contributing to the advancement of scientific knowledge and technological innovations.

1.2.6 Cost savings

Achieving cost savings in research and development (R&D) is a priority for many organizations, as R&D activities can be resource intensive. It is essential for optimizing resources, maximizing efficiency, and ensuring sustainable innovation. One of the means could be the identification and elimination of inefficiencies in R&D processes. Streamlining workflows, reducing unnecessary steps, and optimizing resource allocation can lead to significant cost savings. Lab automation is indeed a strategy that can contribute significantly to cost savings in R&D. While setting up automated systems may require an initial investment, the long-term cost savings can be significant. Automation reduces the need for manual labour and minimizes the consumption of reagents and materials. Its implementation requires careful planning, investment in equipment and technology, training of personnel, and ongoing maintenance. Organizations need to consider the specific needs of their R&D processes and balance automation with other strategies to achieve optimal cost savings and research outcomes.

1.2.7 Safety

Lab automation can significantly increase safety in laboratory environments. Automation technologies are designed to minimize or eliminate various risks associated with manual laboratory processes, thereby enhancing the overall safety of researchers, technicians, and the surrounding environment. For example, reduced human interaction with hazardous materials considered toxic, mutagenic, cancerogenic and irritating, automation minimizes direct human contact with these substances, reducing the risk of exposure and contamination. Automated systems handle samples, reagents, and chemicals with precision, reducing the likelihood of accidental spills, leaks, and releases that could pose safety hazards. Lab automation is also designed to prevent cross-contamination between samples, ensuring that hazardous or infectious materials are contained and isolated. Automation can decrease the need for researchers to wear extensive "personal protective equipment (PPE)", making it easier to follow safety protocols and reducing the likelihood of human error due to discomfort or restricted movement. While lab automation can significantly improve safety, it is important to note that the design, implementation, and operation of automated systems should be approached with careful consideration of safety guidelines and regulations. Proper training and maintenance are also crucial to ensuring the continued safety of personnel and the environment in automated laboratory settings.

1.2.8 Integration and connectivity

Integration and connectivity play a crucial role in the effectiveness and efficiency of lab automation. Integration of various instruments, devices, and software platforms allows for seamless movement of samples and data between different steps of the workflow. This improves overall efficiency, reduces the risk of errors, and accelerates the pace of research and analysis. Automation systems with robust connectivity features enable easy sharing and collaboration of data within the lab and even across different locations. Researchers can access and share experimental data, results, and protocols, facilitating better communication and cooperation among team members. Integration and connectivity allow lab personnel to remotely monitor and control automated processes. This is particularly important for experiments that require constant monitoring or adjustments. Researchers can access real-time data, receive alerts, and make necessary changes without being physically present in the lab. Automation systems often come with built-in data logging and traceability features. Integrating various components of the lab setup ensures that data is accurately captured and attributed to the right source. This traceability is important for maintaining data integrity, complying with regulations, and ensuring the reproducibility of experiments. In addition, automation systems can be integrated with data analysis and reporting tools. This integration facilitates the efficient extraction of insights from large datasets, speeding up the decision-making process and enabling researchers to focus on interpreting results rather than data processing.

1.2.9 Serendipity

Serendipity refers to the unexpected discovery of valuable insights, knowledge, or results while pursuing a different research goal or conducting routine experiments. While lab automation is often associated with planned and controlled experiments, it can still play a role in facilitating serendipitous discoveries. Lab automation can significantly increase the speed at which experiments are conducted. Researchers can run more experiments in a shorter amount of time, increasing the chances of encountering unexpected outcomes or phenomena. Automation allows the researcher to think out of the box, and to try out ideas, and experiments that otherwise would not have been conducted. The Edisonian approach, also known as accelerated serendipity, to innovation is characterized by trial-and-error discovery rather than a systematic theoretical approach. Thus, the accelerated pace of experimentation provides more opportunities for serendipitous discoveries. Automation allows for high-throughput screening of many variables, compounds, or conditions. This process can reveal unexpected correlations, interactions, or effects that might not have been anticipated. These discoveries can lead to new research directions and unexpected breakthroughs. Automated systems generate a substantial amount of data. While the initial goal might be to analyze specific parameters, researchers can also explore the data for patterns, anomalies, or interesting observations that were not part of the original research plan. While lab automation can certainly enhance the potential for serendipitous discoveries, it is important to note that serendipity often relies on the open-mindedness and creativity of researchers to recognize and interpret unexpected results. Automation can provide tools and opportunities, but the human element of curiosity and critical thinking remains essential for transforming unexpected observations into valuable insights.

1.2.10 Flexibility and customization

Lab automation can have both positive and potentially limiting impacts on the flexibility and customization of experiments. Many automated systems are designed to be modular and easily configurable. This flexibility allows laboratories to adapt the automation to their specific needs and accommodate various types of experiments. Lab automation can improve the reproducibility of experiments by precisely controlling variables and minimizing human error. This reliability is particularly important when aiming to replicate experiments for validation or further investigation. It also enforces standard protocols, reducing variability between experiments. This is beneficial when consistent experimental conditions are required for comparisons, quality control, or regulatory compliance. Depending on the design of lab automation systems, some automation systems are optimized for specific protocols and may not easily accommodate deviations or customizations. This can limit researchers' ability to explore novel experimental setups.

1.2.11 Addressing labour shortages

Laboratory 4.0 can contribute to overcoming labour shortages representing the integration of digital technologies, automation, data exchange, and advanced analytics into various industries, including laboratories and research facilities. Lab 4.0 technologies enable the automation of repetitive and manual tasks in laboratories, reducing the need either for many skilled personnel or compensating recruiting shortages. Robots and automated systems can handle tasks like sample preparation and formulation including dispensing steps, application and testing, data collection, and basic experiments, freeing up the limited resources of skilled researchers for more complex and interesting tasks. Laboratories and research facilities that adopt cutting-edge Lab 4.0 technologies may attract a younger and tech-savvy workforce. This can help address labour shortages by appealing to a new generation of researchers. While Lab 4.0 technologies can contribute to overcoming labour shortages in laboratories, it is important to note that their successful implementation requires careful planning, investment, and consideration of the specific needs and challenges of your organization. The early involvement of the dedicated personnel for the future operation of the lab automation systems, starting from planning and designing, setting up, installation, training and start of operation, is key for a successful implementation. Additionally, a balanced approach that combines technological advancements with human expertise is often the most effective way to achieve optimal results.

1.2.12 Competitiveness

Differentiation versus competition can be manifold but needs to provide unique selling propositions. Lab automation can significantly increase competitiveness for a variety of industries and research fields. Industries in specialty chemicals and the formulating industries following in the value chain can benefit from faster development and validation of new products due to automated processes. This shortens the time it takes to bring products to market. Lab 4.0 allows faster and better mapping of the raw material **and** the process space, significantly reduces the cost per experiment and therefore helps to stay competitive. Typically, front runners are more successful by leading the market and increasing their market share. Laboratories that adopt advanced automation technologies differentiate themselves from competitors by demonstrating their commitment to innovation, precision, and efficiency. This can attract clients, collaborators, and investors. It is important to have a clear understanding of the laboratory's specific needs and goals. Not all tasks are suitable for automation, so a thoughtful approach is necessary to identify the processes that will benefit the most. Additionally, ongoing training and support for staff using automated systems are crucial to ensure their effective utilization.

1.2.13 Reproducibility

One of the biggest advantages when using automation in laboratory processes is the reproducibility which often becomes higher. Automation can significantly reduce the variability introduced by human factors, leading to more consistent and reliable results. Automation minimizes the risk of errors that can occur during manual tasks, such as pipetting inaccuracies, measurement mistakes, and data entry errors. Automated systems follow precise instructions, leading to more accurate and consistent outcomes. Automated processes follow predefined protocols and procedures consistently, eliminating variations caused by differences in human technique or interpretation. This consistency enhances the reproducibility of results. Automated processes can be standardized across multiple laboratories or sites, ensuring that experiments are conducted using the same methods and conditions. This promotes consistency and facilitates cross-site comparisons. Increase experimental quality and reproducibility, conserve knowledge in an electronic executable instrument and data files. While automation can significantly enhance reproducibility, it is important to note that proper setup, calibration, maintenance, and validation of automated systems are essential to achieving the desired outcomes. Additionally, automated processes should be thoroughly documented to ensure transparency and reproducibility across different researchers and studies.

1.2.14 Systematics

In the context of laboratory automation, "systematics" refers to the organized and structured approach to designing, implementing, and managing automated processes and systems within a laboratory environment, which builds most probably the fundament for the investment into Lab 4.0. It involves careful planning, coordination, and consideration of various factors to ensure the successful integration and operation of automated solutions. This can be broken down into several elements as listed below.

- Planning and assessment
  - Identify the specific tasks or processes that can benefit from automation.
  - Define clear goals and objectives for implementing automation, such as increased efficiency, improved accuracy, or enhanced reproducibility.
  - Assess the feasibility of automation for each identified process, considering technical requirements, costs, and potential benefits.
- Process analysis
  - Break down the workflow into individual steps and tasks.
  - Analyze each step to understand its requirements, dependencies, and potential points of variation.
  - Identify potential bottlenecks, sources of error, and areas where automation could have the most impact.
- Technology selection
  - Research and select appropriate automation technologies, instruments, and software platforms based on the specific needs of the laboratory.
  - Consider factors such as throughput, precision, compatibility with existing systems, and scalability.
- Integration
  - Ensure that the chosen automation solutions can seamlessly integrate with existing laboratory equipment and software.
  - Address compatibility issues and potential challenges related to data exchange and communication between different systems.
- Customization and programming
  - Customize the automation workflow to match the laboratory's unique requirements.
  - Develop programming scripts or protocols that dictate how the automated processes will be executed.
  - Consider user-friendly interfaces that allow for easy control and monitoring of the automated systems.
- Validation and quality control
  - Establish validation protocols to ensure that the automated processes produce accurate and reliable results.
  - Conduct thorough testing to verify the performance of the automated system under different conditions.
  - Implement quality control measures to detect and address any deviations from expected outcomes.
- Training and documentation
  - Train laboratory personnel on how to operate, troubleshoot, and maintain the automated systems.
  - Create comprehensive documentation that outlines the operating procedures, maintenance schedules, and troubleshooting steps.
- Data management
  - Implement data management protocols to handle data generated by automated processes.
  - Ensure proper data storage, retrieval, and analysis to support research and decision-making.
- Change management
  - Prepare for changes in workflows and processes due to the introduction of automation.
  - Address any resistance or concerns among staff members through effective communication and training.

- Continuous improvement
  - Regularly assess the performance of the automated processes and identify areas for improvement.
  - Implement updates, optimizations, and enhancements to keep the automation system aligned with changing needs.
- Sustainability
  - Consider the long-term sustainability of the automation solution, including factors like maintenance, upgrades, and scalability.

Systematics in lab automation involves a holistic approach that encompasses technical, operational, and organizational considerations. It aims to maximize the benefits of automation while minimizing potential challenges and disruptions. Properly implemented systematics can lead to more efficient, accurate, and reliable laboratory operations.

1.3 Summary

Overall, laboratory automation offers numerous advantages that can lead to increased efficiency, data quality, and scientific advancements in various research and industrial applications. It complements the work of researchers and plays a crucial role in modern laboratories, see Figure 1.5. High throughput (HT) originally stems from the pharmaceutical industry. However, over the years the term got stretched as general term for automation which is not anymore according to its origin for yes / no screening, so basically number of experiments driven. High output (HO) aims at medium number of experiments with as much information as possible (temperature, pressure, mixing speed, feed rates, ...), but not just information, information which is scalable. High throughput experimentation (HTE) or shotgun-approach, is to «shoot» in the entire space of interest, providing leads/hits/candidates. High output experimentation (HOE) is directed educated research, providing

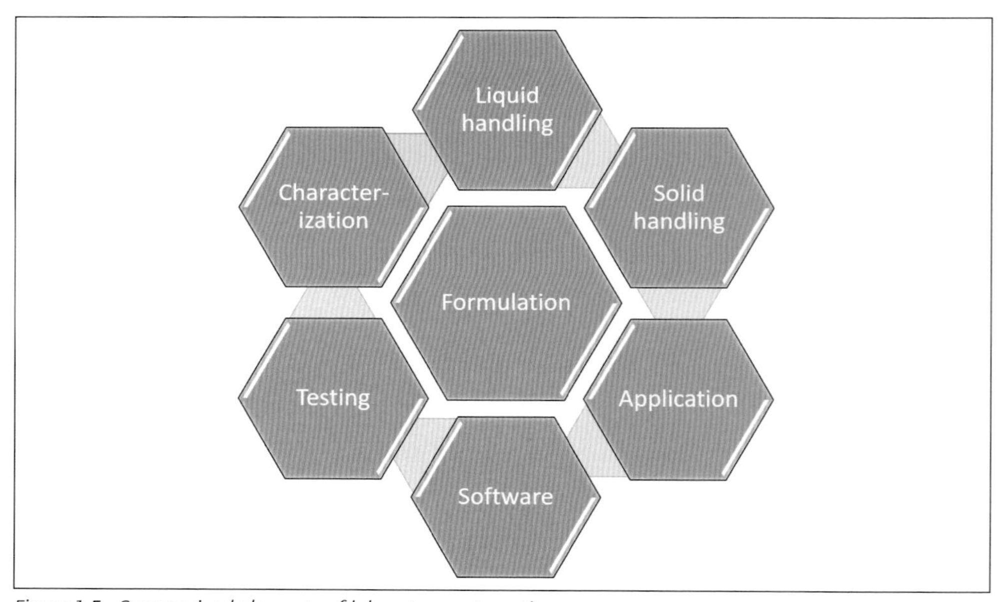

Figure 1.5: Summarized elements of laboratory automation Source: PERFECO Consulting Gysau

scalable data to knowledge, moving library by library through the information space, finally resulting in finding one of the optima. If the local optimum is not fulfilling the project requirements, the directed educated approach is simply repeated with another lead/hit/candidate.

1.4 References

[1] HALL, Ch., A History of Paint (Part One), Burnaway Digital Magazine, 2015, https://burnaway.org/magazine/a-history-of-paint-part-one/

[2] OLSEN, K. (1997) Rosie the Robot, Laboratory Automation and the Second World War, 1941 to 1945. Lab. Robot. Autom., 9(3), 105-112.

[3] FERGUSON, B. (1942) Semiautomatic Fractionation: A Rapid Analytical Method. Ind. Eng. Chem. Anal. Ed., 14(6), 493.

[4] GILSON, R., GILSON, W. (1974) Adjustible Pipett. U.S. Patent 3827305.

[5] FELDER, Robin A. (2006-04-01). "The Clinical Chemist: Masahide Sasaki, MD, PhD (August 27, 1933 – September 23, 2005)". Clinical Chemistry. 52 (4): 791–792.

[6] BOYD, J. (2002-01-18). "Robotic Laboratory Automation". Science. 295 (5554): 517–518.

[7] LIM Source, a laboratory information management systems resource". Archived from the original on 2009-08-11. Retrieved 2009-02-20.

[8] HAMILTON, S.D. (1986) Robotic assays for fermentation products. In: Advances in Laboratory Automation Robotics, Vol. 4 (Strimatis, J. and Hawk, G. L., eds.), Zymark (Hopkinton, MA), 1-23

[9] AKOWITZ, A. (1997) High Throughput Screening: A Simplified Protocol to Verify Pipetting Accuracy. Laboratory Automation News, 2(2), 33-35.

2 Path to automation

The reasons for starting automation can have different causes. The drivers for automation and the benefits of automation are two different aspects, but they are closely linked. Overall, the drivers for automation are the factors that motivate companies to implement automated solutions. These are, for example, market trends and requirements, technological advances that facilitate investment, regulatory requirements or cost pressure. The benefits of automation and the positive effects that these solutions can have on companies, such as increased efficiency, cost savings and quality improvement. The drivers and benefits are presented in much greater detail below. What is striking here is the close connection between the individual considerations, which also reinforces the decision for a possible investment.

2.1 Drivers for automation

Many processes have changed with the onset of industrialization. This affects not only the traditional production of goods, but also many other disciplines such as research and development. What they have in common is an increase in efficiency combined with greater speed. The same goals also apply to the research and development of paints and varnishes and other types of coating materials. The demand for the development of products with better performance, less harmful ingredients for users and consumers, reduced ecological footprint, use of sustainable raw materials and secondary influences such as legislation, social demands, market and competitive situation, cost and price pressure, shortage of skilled workers can very often no longer be met with existing resources. The pressure is constantly increasing, forcing companies to cut back or feverishly search for ways out. So what are the real drivers for automation?

2.1.1 Globalization

Probably the biggest driver of the economy and global change was standardization in logistics. Since the introduction of sea containers in 1956 by the American Malcom McClean [1], world trade has rapidly gained momentum. This also marked the beginning of the globalization of raw materials and the products formulated from them. In order to survive this rapid growth, more and more innovations had to be developed more quickly, and the resulting products successfully launched on international markets. In addition to product development, the demand for services has also increased, particularly in the area of application technology. Companies that have missed out on this development have been and will sooner or later be strategically acquired by stronger competitors.

2.1.2 Digitalization

As a result of rapid global growth, the exchange of information and data of all kinds has also picked up speed. The introduction of microcomputers was the beginning of storing data digitally, but no more and no less. As before, much data was then processed in analog form, for example from printed reports and tables of entered laboratory results, etc. Further progress in accelerating innovation came with the introduction of statistical experimental design. Here too, microcomputer technology helped to speed up the planning and evaluation of experiments. The further development of software applications made many measuring instruments and analytical devices more powerful, particularly in terms of operation, but also in the evaluation and presentation of results. At the same time, the exchange of electronic data by e-mail has also become easier. Solutions were also developed to manage increasingly large volumes of data and exchange it via cloud solutions. In addition to traditional data processing, the need for characterization using image and video files has increased. The ever-improving and more demanding need for precise characterization and evaluation of data of all kinds can no longer be met by manual processes. New software solutions have come to the rescue and the latest developments in the introduction and use of artificial intelligence [2] and machine learning [3] currently pose the greatest challenges for data processing. However, immense amounts of data are required for this to be used effectively. The larger the data pool, the better the results provided by AI and ML. The fastest method for generating gigantic amounts of data is laboratory automation with high-throughput systems.

2.1.3 Increasing legislation

Great and ever-increasing importance is attached to compliance with legal regulations. In particular, the safe handling of chemicals is leading to a dynamic environment in legislation due to increasingly precise laboratory methods and lower detection limits. The classification of chemicals and their use and development of preparations is constantly being adapted and documented [4]. It is also quite right that people, whether scientific staff, application technicians or end users, to name just a few examples, should know how to handle the relevant chemicals and what personal protective measures should be used to ensure hazard-free handling. For the European member states, the European Regulation on Registration, Evaluation, Authorization and Restriction of Chemicals (REACH) and the Classification, Labeling and Packaging of Chemicals (CLP), for example, provide information on this. This provides information. Specific industries such as the pharmaceutical, food and healthcare sectors require strict quality control and regulatory compliance. Digitization and standardization of QC processes can ensure compliance and avoid costly issues with the authorities. Regardless of the industry, inspections and their documentation are constantly increasing. In order to cope with this volume and its dynamics, it is obvious to use the help of laboratory automation for this as well. In addition, the use of laboratory automation greatly reduces the exposure of employees to chemicals, particularly hazardous chemicals, and thus actively contributes to the protection of people.

2.1.4 Cost reduction

In every company, a positive operating result is crucial for sustainable development and continued existence. This is in the interests of all stakeholders, not just the shareholders. One option is to pass on costs directly to customers. However, this does not usually contribute to competitiveness in the long term. The alternative is to save costs in the selection of raw materials and their suppliers, cost-conscious product development, manufacturing processes and the marketing of products. The use of laboratory automation is suitable for the cost-efficient development of paints and paint formulations without sacrificing innovation. In particular, the screening of raw materials from different manufacturers and the quantities used can be carried out on a much larger scale using high-throughput systems. The mass of results gives the developer security in terms of performance and, thanks to a significantly reduced development time, leads to a faster market launch and securing, and possibly expanding, competitiveness. Time to market is the keyword here. In addition to R&D, standardization and the use of automation can also lead to cost savings in quality control (QC) by increasing efficiency, reducing errors and optimizing resource allocation [5]. Examples include coloristics in production for monitoring the manufactured color shades, but also other properties such as flow behavior (viscosity, rheology), solids content, specific weight and much more.

2.1.5 Explosion of accumulated data

As a result of the introduction of initial automation in the area of R&D, the amount of data is also increasing. The large amount of data generated, for example due to more efficient or faster manual manufacturing processes in the area of raw material screening, but also the use of semi-automated analysis devices by means of autosamplers, requires more time for their evaluation. The targeted use of software and thus the start or increase of data digitization can increase efficiency here. The amount of data generated in research and development is often growing exponentially. Digitization therefore helps with the effective management and analysis of this data and facilitates data-driven decision-making. Nevertheless, the amount of data obtained can often still be insufficient for the use of more advanced software solutions in the field of AI. This requires the expansion of existing laboratory automation in extremis to high-throughput systems in order to fully utilize AI and ML.

2.1.6 Customer expectations

Today's customers expect high-quality products and fast delivery. Accelerating R&D and maintaining quality standards are critical to meeting customer requirements. This applies to both the raw materials industry and the formulating industry. In the face of tough competition, raw material manufacturers carry out series tests to develop and continuously optimize raw materials and preparations in order to meet the expectations of paint and coating producers. As usual, the criteria are costs/prices, safety in terms of handling and compliance with legislation with few to no restrictions, availability and, of course, performance.

A similar picture applies to manufacturers of paints and coatings. However, their efforts are focused on applicator properties, i.e. easy handling with the best possible performance at a competitive price. All of these properties in combination are in themselves a contradiction in terms, as there is no such thing as a one-size-fits-all solution.

That is why it is important to explore the maximum without increasing the costs immeasurably. This again speaks in favor of investing in laboratory automation, taking into account the return on investment.

2.1.7 Risk reduction

Standardizing QC processes and digitizing records can help identify and fix problems in real time, reducing the risk of costly product recalls. Wherever paints and coatings perform critical functions, there is also a guarantee claim behind them. A malfunction in electrical insulation paints, an insufficiently reflective paint or the unintentional use of hazardous raw materials in the paintwork of toys or in the food sector, for example, can lead to damage to people's health or very high property damage. The selected quality tests and their frequency must be weighed up. These can be randomly selected or defined random samples or, depending on the risk potential, the testing of every batch. The decision for this is once again based on risk assessment and testing capacities. The use of laboratory automation for quality control can eliminate the bottleneck here.

In addition to reducing the risk of potential quality and performance errors, automated laboratory systems protect employees from carrying out hazardous process steps and handling CMR substances that are carcinogenic (C), mutagenic (M) and reprotoxic (R) [6]. In addition to the actual contact, the exposure times over a longer period of time are also greatly reduced or, in the best case, eliminated.

The COVID-19 pandemic was an impressive example of maintaining innovative strength through laboratory automation in times of crisis. Systemically relevant activities and functions, for example in the healthcare sector, were maintained despite the risk of infection and the spread of viruses. On the other hand, non-systemically relevant activities were massively reduced. The operation of a laboratory automation system requires only a small number of staff and enables major and rapid research progress compared to most competitors, despite the absence of employees, most of whom are working from home.

2.1.8 Standardization, quality & reproducibility

One of the essential and widely known features of automation is the repetition of predefined processes. Execution is repeated with high precision and reliability until the job is completed. As a result, the quality is always comparable, and the reproduction rate is extremely high. Automation is therefore particularly suitable for the standardization of processes, regardless of whether they are carried out at the same or different locations. This is particularly advantageous for internationally active companies. This guarantees objectivity, especially when it comes to quality checks despite different production sites. Objectivity is guaranteed with automation, regardless of the time

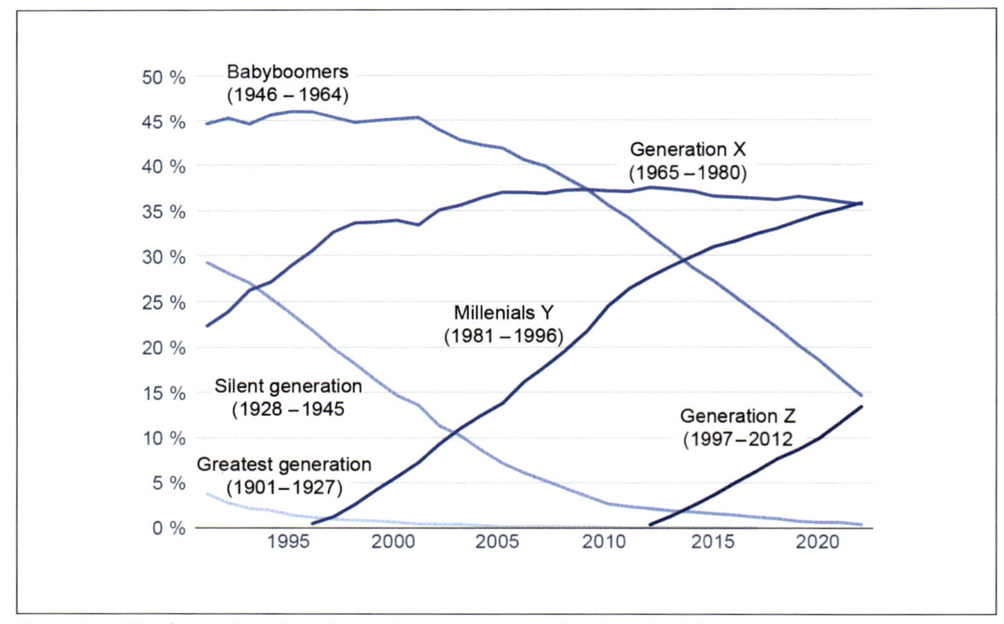

Figure 2.1: The figure describes the working population of Switzerland from 1991 to 2022

Source: Swiss federal office for statistics

and day of the week. Automation is also unaware of any variations in daily form, as is absolutely natural with humans. Furthermore, automation does not know fatigue, weekends, sick days and vacations, which makes capacity utilization and order execution more reliable in terms of planning.

2.1.9 Competitiveness & productivity

There is no doubt that the speed of high-throughput systems, combined with the possibility of 24/7 operation, is an outstanding argument and therefore a driver for investing in laboratory automation systems. Compared to manual processes, this leads to significantly increased productivity and ultimately to a decisive competitive advantage. In the area of raw material development for the paint and coatings industry, experience shows an increase in productivity by a factor of 5 to 80 [7]. The differences are essentially based on the complexity of the workflows selected for automation. The frontrunners of innovations usually secure the market. The followers, depending on their late market entry, will rarely be able to catch up in terms of time or technology. Consequently, the introduction of automation in R&D is a sustainable and forward-looking step towards securing a successful market. The faster and stronger innovative strength amortizes the initially seemingly high investment costs within a short period of time.

2.1.10 Solving the shortage of skilled workers

A relatively new argument for the automation of manual laboratory processes is the increasing shortage of skilled workers. In highly developed economies in particular, society is ageing and the number of people applying for jobs is decreasing proportionately [8], see Figure 2.1.
The baby boomers are retiring and the number of people in employment, i.e. the next generation of skilled workers from generations X, Millennials and Y, is not able to compensate for this. A logical consequence of this shortage of skilled workers is investment in laboratory automation to ensure the innovative strength and continued existence of companies.

2.1.11 Data integrity

Data is now regarded as the most important asset of companies, even in laboratory areas. In the course of digitalization, this great importance will continue to increase. Data integrity refers to the accuracy, completeness and reliability of data throughout its lifecycle. It is critical that data remains accurate and unaltered, both during its creation and during its storage, transmission and processing. A lack of data integrity can lead to inaccurate information, unreliable analysis and potentially serious consequences, such as mis-development, labelling of preparations or reputational damage.

In the scientific debate on the subject, data integrity is divided into data consistency, data security and data protection.

- The term data consistency refers to the logical correctness of the data or "semantic integrity". Translated, this means that the data entered must comply with the previously defined rules.
- Data security is about ensuring that the data is physically protected – against loss, damage and unauthorized access. Dangers here include operating errors or technical errors. The aim is to continuously back up the data.
- The aim of data protection is to prevent the misuse of data. The aim is to protect personal data and thus safeguard privacy. Above all, technical and organizational processes are required.

The most important prerequisite is followed by the most important measure – the introduction of comprehensive database management. This ensures data integrity throughout the entire life cycle. On the one hand, the ever-increasing technology surrounding databases is causing problems, while on the other, an increasing flood of new data can be observed. Fortunately, there are ways and means of simplifying the complex processes of database management. Namely by automating routine tasks to a large extent. These include, for example, script execution or status checks.

2.1.12 Artificial intelligence & machine learning

Artificial intelligence (AI) is definitely a driving factor for automation. AI can automate repetitive tasks that would otherwise require a lot of time and resources. By automating these tasks, companies can increase their efficiency and free up employees for value-adding activities. Furthermore, improved decision-making is possible due to the analysis of large amounts of data by AI systems, the recognition of patterns and the making of predictions, an increase in accuracy and responsiveness in real time. By integrating AI into various development processes, laboratory staff can optimize workflows, identify bottlenecks and automatically improve or redesign inefficient processes. In addition, AI can be used to analyze the behaviour of raw materials and formulations, but also of users.

Automation through AI can lead to significant cost savings as it reduces the need for human labour in certain areas or increasing the efficiency of human workers. It can also lead to a reduction in raw material costs linked to performance parameters. Additionally, scalability can be increased with AI systems to handle increasing data and workloads without the need for additional human resources. This enables companies to keep pace with growth and respond flexibly to changing requirements.

Overall, AI plays a crucial role in transforming businesses and creating smarter, more efficient and agile organizations through automation. This applies in particular to research and development, since very large amounts of data are also processed here. The use of high-throughput systems, which generate a constantly and rapidly growing data pool for AI in the first place, is a major advantage here.

The same is true for machine learning (ML). Overall, it plays a central role in the automation of processes and functions. Companies can benefit from ML-driven automation solutions by increasing efficiency, reducing costs and improving the user experience. ML drives the automation of decision-making processes, process optimization, prediction and prevention as well as automated data collection and processing. This in turn leads to time and cost savings due to the faster and more efficient execution of tasks.

2.1.13 Elimination of human error sources

It is generally assumed that automation eliminates human error. Automation can help to reduce human error, but it cannot eliminate it completely. This is relatively easy to explain.

Automation systems are only as good as their programming and configuration. If errors occur in the automation processes, they can have serious consequences. These errors can be caused by human errors in programming, insufficient data quality or unforeseen events. This circumstance is especially heightened in complex environments, as automation systems may not cover all scenarios and constraints. Without human intuition and flexibility, unexpected situations often cannot be handled or decisions made in unclear situations.

Even though many tasks can be automated, many processes still require human supervision, control and intervention. Humans often need to intervene to handle exceptions, solve complex problems or make strategic decisions.

Although automation cannot completely eliminate human errors, it can help to reduce their frequency and severity. By combining automation with appropriate monitoring, testing and error correction mechanisms, organizations can minimize the risk of human error while reaping the benefits of automation.

2.2 Benefits of automation

2.2.1 Faster development time to market launch

Laboratory automation can significantly shorten the development time to market launch. There are many reasons for this. The first is that automated laboratory equipment and robotic systems perform experiments faster than manual processes. The factor for accelerated experiments can vary from system to system. A typical value is between a factor of 10 and 20 but can also assume values of over 100. It must be taken into account that automation systems can in principle enable 7/24 operation. In addition, robotic systems do not have any recovery phases as is the case with humans, e.g. weekends, vacations, public holidays and 8-hour working days. This enables researchers to carry out more experiments in less time and collect data more quickly.

Automation reduces the likelihood of human error and increases the consistency and reliability of experimental results. This contributes to the collected data being more accurate and reproducible. Parallel processing is also advantageous. Automation allows several experiments to be carried out simultaneously or in parallel steps. This enables a more efficient use of resources and speeds up the development process.

The creation of optimized workflows for automated laboratory equipment can simplify the execution of complex experiments and analyses. This helps to avoid bottlenecks and speed up the overall process. Subsequently, by integrating fast data analysis with data analysis technologies such as artificial intelligence and machine learning, it is possible to process and interpret data faster. The shortened experimentation time and improved data quality provide faster feedback and enable iteration cycles to be shortened, see Figure 2.2.

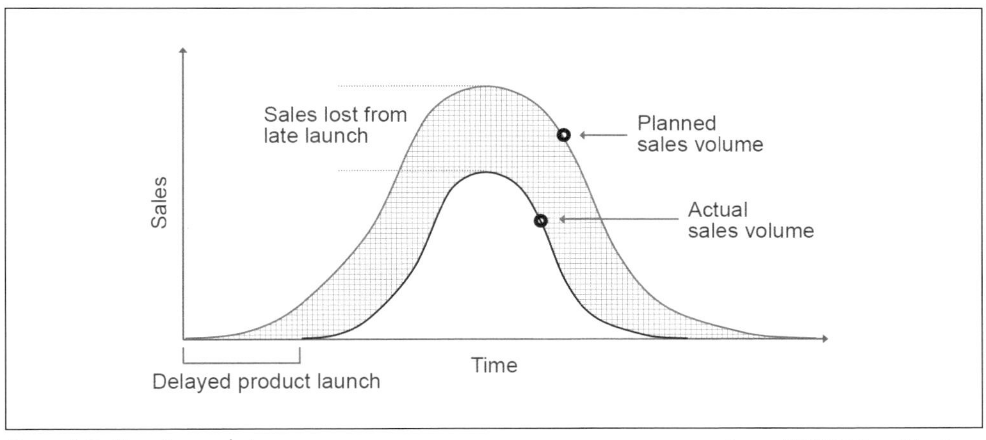

Figure 2.2: Time to market *Source: PERFECO Consulting Gysau*

Overall, laboratory automation helps to speed up the development process by conducting experiments faster, processing and analyzing data faster and improving overall efficiency. This enables companies to develop and bring products to market faster, which provides a competitive advantage, especially in dynamic industries.

2.2.2 Improved product quality

It is obvious that experiments can be carried out with greater precision and reproducibility by automating laboratory processes. High-precision dosing and measuring systems significantly increase the precision and miniaturization of laboratory experiments. This reduces variation between experiments and leads to more consistent results, resulting in improved product quality.

Automated lab equipment can help minimize human error and improve control over experimental variables. This allows researchers to collect more accurate data and minimize potential confounding factors. In addition, automated laboratory systems can be equipped with sensors and monitoring functions that continuously monitor process parameters and provide early warning of deviations.

In the area of quality control, better compliance with quality standards is achieved. Through automation, companies can ensure that their production processes meet the required quality standards. This can help improve compliance with regulations and industry standards and minimize the risk of production errors.

Overall, laboratory automation helps to improve product quality by standardizing processes, collecting and analyzing data accurately and enabling efficient error detection and correction.

2.2.3 Cost efficiency

By using digital technologies, R&D and QA processes can be automated and optimized. This reduces manual effort, minimizes errors and speeds up the execution of experiments, tests and analyses, which ultimately leads to cost savings. One of the main advantages of laboratory automation is the cost savings that can be achieved. Automated systems allow a considerably larger number of samples to be processed, significantly reducing the cost per sample.

In addition, digital technologies facilitate collaboration between teams and sites through virtual platforms, cloud-based tools and collaboration software. This enables more efficient communication, knowledge sharing and project management, which in turn leads to improved productivity and cost efficiency.

2.2.4 Better use of data

Automation combined with digitalization enables extensive collection and analysis of data in real time. The creation of data standards is an advantage here. These can help laboratories to solve many problems that have arisen as a result of the enormous growth in knowledge and data. Data standards also help to exchange information more easily, efficiently and transparently. In addition, they make it easier to comply with data principles, which play an important role, especially in meeting regulatory requirements. There are now many organizations and initiatives that are already promoting standards, guidelines and best practices in the field of laboratory and life sciences data [9].

Advanced analytical tools and algorithms can process large amounts of data and provide valuable insights. Particularly in screening, it enables the identification of trends and statements on optimization potential. Accuracy increases as the amount of data increases, delivered through the use of laboratory automation and, in particular, high-throughput systems.

2.2.5 Flexibility and scalability

Crucial features in the implementation of laboratory automation solutions are flexibility and scalability. This is of great importance in a dynamic development environment. For example, the production, application and testing of architectural paints and industrial coatings are very different, despite many similarities. Depending on the configuration of a laboratory automation system, different workflows for paint and coating systems can be taken into account. This usually increases the amount of hardware used, but can be combined with purposefully designed modularity, which can work independently of each other in parallel. In this way, the automation solution can be adapted to specific requirements and, if necessary, additional modules can be added or removed to meet changing requirements. The basic prerequisite for this is a scalable architecture that allows the capacity and performance of the system to be expanded as required.

Due to the high investment costs, it is possible to split the overall automation requirements into separate phases. In this way, the final goal of the desired automation can be tackled in stages and the costs can be controlled over defined investment periods. Of course, this requires a far-sighted and sustainable planning phase in order to avoid investing in isolated stand-alone solutions that can no longer be linked together. The future space and personnel requirements as well as the software solutions for the planned workflows and control of the system must also be taken into account from the outset to ensure that the project is ultimately a success.

2.2.6 Compliance with legal regulations

Laboratory automation can make an important contribution to compliance with legislation in various industries, especially in areas where strict compliance requirements apply. For example, the integrity of data must be guaranteed, and complete traceability of samples and experiments must be ensured. By using electronic laboratory notebooks (ELNs), barcode and RFID technologies as well as automated data collection and processing systems, companies can ensure that all laboratory activities meet the requirements of compliance guidelines.

In addition, laboratory automation can improve compliance with health and safety regulations by minimizing human exposure to hazardous substances and risks. By automating hazardous or high-risk laboratory processes, organizations can reduce the risk of workplace accidents and injuries while ensuring compliance with relevant regulations.

It is also worth mentioning that audit and compliance processes can be made more efficient and the documentation of compliance activities can be facilitated. By providing comprehensive audit trails, reporting capabilities and audit-proof data storage solutions, companies can demonstrate adherence to compliance guidelines and respond to requests from regulatory authorities when necessary.

Overall, laboratory automation helps to improve regulatory compliance across various industries by supporting data integrity, traceability, quality control, health and safety standards, and audit and compliance management. By implementing automated laboratory solutions, companies can minimize the risk of non-compliance and penalties while improving the efficiency, productivity and competitiveness of their laboratory processes.

2.2.7 Increased cooperation

By networking laboratory devices and systems, recurring tasks can be automated, such as sampling, measuring, dosing and mixing. This reduces manual workload, minimizes errors and improves the consistency and accuracy of results. Collaboration between researchers and laboratory teams, regardless of their location, can be facilitated. Furthermore, improved planning and utilization of resources, e.g. laboratory systems, consumables and working time, can be enabled through increased collaboration. By sharing data, protocols and results, researchers can collaborate more effectively, exchange ideas and learn from each other, leading to faster progress and innovative solutions.

Connected labs therefore help to increase efficiency and productivity in laboratory operations by automating processes, optimizing resources, facilitating collaboration and supporting compliance requirements. By integrating digital technologies and networked solutions, companies can become more competitive and respond more quickly to changing requirements and challenges.

2.2.8 Environmental and sustainability goals

A major advantage of laboratory automation is the miniaturization of processes. In manual processes, experiments are usually produced and carried out in significantly larger quantities due to a lack of precision and the associated reproducibility and reliability. Automation makes it possible to significantly reduce the size of experiments and at the same time increase precision. Furthermore, automation allows only as many samples to be produced as are required for testing. This reduces the amount of raw materials used per experiment. During the execution of tests, fewer emissions are produced due to the small quantities. In the best-case scenario, little to no material remains after all tests have been completed, which massively reduces or eliminates the amount of waste. This significantly conserves resources and therefore also the environment.

Automated laboratory systems can help to improve compliance with environmental standards and regulations by enabling comprehensive documentation and traceability of environmental impacts. By automating the collection and reporting of environmental data, companies can better meet regulatory requirements and comply with environmental standards.

By using laboratory automation solutions, various sustainability goals can be achieved, including energy efficiency, waste reduction, water efficiency, emissions reduction, promotion of recycling and circular economy, and social responsibility. This enables companies to operate more sustainably and make a positive contribution to environmental protection and social responsibility.

In summary, laboratory automation can have a very positive and sustainable impact on the environment by contributing to a more efficient use of resources, a reduction in waste and emissions and better compliance with environmental standards. By integrating environmental and sustainability practices into laboratory operations, companies can contribute to environmental protection and develop long-term sustainable business models.

2.2.9 Competitive advantage

Companies, especially the early adapters, can secure decisive competitive advantages through the numerous benefits of laboratory automation. This is achieved through
- increased efficiency and productivity
- consistent and reliable results
- time and cost savings
- faster time-to-market
- more capacity and scalability
- standardization and comparability
- better use of specialist knowledge

By reducing manual working hours, optimizing the use of resources and avoiding errors, companies can lower their operating costs and strengthen their competitiveness. In addition, laboratory automation allows a company to react much more agilely and adapt to changing market requirements. By automating routine and repetitive tasks, companies can use the expertise of their employees for more demanding and strategic tasks. This helps to increase employee commitment and satisfaction and make the company more competitive overall. A company that invests in innovative, future-oriented laboratory automation also increases its attractiveness and reduces staff turnover and thus the loss of knowledge.

2.2.10 Regulation and monitoring

Regulatory and process monitoring requirements can be more easily achieved through the use of laboratory automation. Comprehensive documentation and traceability of laboratory activities is provided through the automated capture of data, logs and results, and organizations can accurately document who performed which tasks, when they were performed and under what conditions, facilitating compliance with regulatory requirements.

Laboratory automation solutions can integrate compliance checks and alerts to ensure that all activities comply with applicable regulations. By implementing automated monitoring systems, companies can ensure that all processes and procedures meet the relevant compliance requirements and intervene in good time in the event of deviations. At the same time, they help to improve quality control and assurance by providing precise and reproducible results. By automating tests, analyses and inspections, companies can ensure that products and processes meet the required quality standards and specifications, helping to comply with regulatory requirements. In addition, real-time monitoring of laboratory activities and automated reporting of results is enabled. This facilitates the timely identification of problems or deviations from regulations and enables companies to react quickly and take appropriate corrective or preventive action.

2.2.11 Innovation & creativity

Promoting interdisciplinary collaboration is an often underestimated aspect of laboratory automation. It can promote collaboration and knowledge sharing between different disciplines and teams. By integrating data, protocols and resources, researchers from different fields can work together on innovative projects and learn from each other. Increasing laboratory throughput and capacity allows more experiments to be conducted simultaneously and a wider range of ideas and approaches to be explored. Automated laboratory equipment and systems cannot always run processes faster than is possible manually, but these are more in total as the automation also works at night and on weekends. If several process steps are carried out simultaneously, this results in a further multiplier compared to manual processes.

The advantage is that large volumes of data are generated, which can then be evaluated using big data analysis techniques and advanced data analysis. This enables researchers to identify patterns, trends and correlations that can lead to new insights and innovative solutions. Experiments and results are easier to reproduce and validate, which increases the credibility and trustworthiness of scientific staff.

Laboratory automation thus helps to promote innovation and creativity by providing researchers with more time and resources for creative tasks, improving the quality and consistency of data and enabling access to large amounts of data.

2.2.12 Adoption of new technologies

The introduction or expansion of laboratory automation accelerates the adoption of new technologies. Methods that have been used for decades in manual laboratory operations are very often not suitable for automation. This is particularly difficult when complex, sometimes almost impossible, efforts have to be made to clean test equipment automatically. This requires a willingness and flexibility to use new technologies and methods. Unfortunately, rigid and not always progressive standards (ISO, ASTM, etc.) all too often stand in the way of this. These are anchored in many regulations and specifications, which is why deviating from them may prevent the introduction of automation. For developers, however, it is crucial that the measured results of old, traditional methods are comparable with the new technologies.

There are many examples of "outdated" methods, which is why only one method is mentioned here as an example. The direct measurement of the density of liquids using a pycnometer in accordance with ISO 2811-1 or ASTM D 1475, see Figures 2.3 and 2.4. The pycnometer is filled with the sample to be analyzed in a balanced and bubble-free manner and tempered to 20 ± 0.5 °C. The pycnometer is then placed on the analytical balance and weighed. The process requires a cleaning step of the leaked material through the hole in the lid before weighing. After weighing, the pycnometer must be emptied and cleaned without leaving any residue. In addition to these steps, which are not automation-friendly, a large amount of material is also required - at least 50 ml, 100 ml as standard. This could be achieved with a significantly smaller sample quantity using modern density meters, see also Chapter 7.1.6 Density.

Figure 2.3: Pycnometer ISO 2811 with 100 ml Source: Byk-Gardner GmbH

Figure 2.4: Pycnometer ASTM D 1475 with 0.0220 gallon or 83.2 ml Source: Byk-Gardner GmbH

In addition to new hardware technologies, there are also significant innovations in the area of software. The applications of design of experiments (DOE), artificial intelligence (AI) and machine learning (ML) are already on everyone's lips but are often only at the beginning of their use, especially for the two software solutions mentioned last. Their meaningful use first requires the availability of very, very large amounts of data. This is precisely why the use of laboratory automation or high throughput experimentation (HTE) makes sense.

2.2.13 Serendipity

Laboratory automation enables the use of high-throughput technologies that allow researchers to quickly generate and process large amounts of data, perform exploratory experiments and pursue unconventional approaches. By automating high-throughput techniques such as screening methods or sequencing technologies, researchers can work faster and more efficiently and potentially make serendipitous discoveries.

HTE provides researchers with more time, resources and flexibility for creative and exploratory activities, enables broader data collection and analysis, increases flexibility for exploratory experiments and assists in hypothesis generation. This helps to accelerate the research process and increase the likelihood of serendipitous discoveries.

2.3 Literature

[1] LEVINSON, M. (2016, 2nd ed.). – The Box: How the Shipping Container Made the World Smaller and the World Economy Bigger. – Princeton, New Jersey: Princeton University Press.

[2] NAUGLER, C.; CHURCH, D. L. Automation and artificial intelligence in the clinical laboratory. Critical reviews in clinical laboratory sciences, 2019, 56. Jg., No. 2, p. 98–110.

[3] FORD, B. A.; MCELVANIA, Erin. Machine learning takes laboratory automation to the next level. Journal of clinical microbiology, 2020, 58. Jg., No. 4, p. 10.1128/jcm. 00012-20.

[4] www.consilium.europa.eu/en/policies/chemicals/

[5] KRIEGBAUM, E. Interview: "We see a need to catch up". European Coatings Journal, 2022, Nr. 12, S. 30.

[6] GENZEN, J. R.; BURNHAM, C. D.; FELDER, R. A.; HAWKER, C. D.; LIPPI, G.; PECK PALMER, O. M. (2018) Challenges and Opportunities in implementing total laboratory automation. Clin. Chem., 64, 259-264

[7] RECKTER, B. "Innovation hat auch mit Mut zu tun". VDI News, 2022, Vol. 1, p. 18

[8] Working population of Switzerland from 1991-2022, Swiss Federal Office of Statistics, www.bfs.admin.ch/bfs/de/home/statistiken/arbeit-erwerb/erwerbstaetigkeit-arbeitszeit/alter-generationen-pensionierung-gesundheit/generationen-arbeitsmarkt.html, published on 20th Feb 2023

[9] HOLLMANN S, KREMER A, BAEBLER Š, TREFOIS C, GRUDEN K, RUDNICKI WR, TONG W, GRUCA A, BONGCAM-RUDLOFF E, EVELO CT, NECHYPORENKO A, FROHME M, ŠAFRANEK D, REGIERER B, D'ELIA D. The need for standardisation in life science research - an approach to excellence and trust. F1000Res. 2020 Dec 4;9:1398.

3 Automation solutions

The chemical industry, and therefore also the paint and coatings industry, is under constant pressure to reduce its actual costs for tests and analyses and at the same time comply with the accelerated and increasingly numerous test plans in order to drive innovation. Driven by new technical requirements, socio-political pressure, stricter regulations and restrictions and even bans on chemicals, raw material availability and rising raw material prices, the number of companies seriously considering the introduction or expansion of laboratory automation is constantly increasing. For this reason, laboratories need to implement tools and equipment that minimize total cost of ownership and maximize process efficiency. Laboratory processes lend themselves to robotic automation because they involve numerous repetitive motions where the steps are repeated the same way every time. Which processes should be automated first? Typical examples are transfer processes of containers and substrates, dosing of raw materials, mixing and dispersing processes, different forms of application, cleaning steps, drying processes, testing and characterization of wet samples and films, stability tests, etc. Repetitive tasks are time-consuming, boring and prone to human error. For them to work, the automated processes must be well understood, standardized, robust and repeatable.

If the lab needs to replace outdated instruments to meet the need for higher throughput, this is the right time to consider a move towards automation. These higher throughput requirements may stem from the need to validate product performance, obtain more research data and perform more test runs to find the most suitable product. Many product innovations have been developed as a result of 'smart failures'. If the lab is able to run more test cycles, the likelihood of breakthrough products being created due to such smart failures increases.

In principle, there are two approaches to laboratory automation. These can be used in isolation or in combination. Automation for research and development has high demands on many flexible workflows, due to the universal use for a wide variety of coating systems and the associated large number of different raw materials. In addition, the type of inspection and test instruments also differ. In the area of quality assurance, significantly fewer different workflows are used, often just one or two workflows. However, the requirements for reproducibility, precision and robustness in continuous operation are even higher.

There are different variants for laboratory automation, some of which are independent of the topics described before. These can relate to the simplest tasks in sample preparation and transfer processes. More advanced are complete workflows that can be implemented in isolation as standalone solutions. It becomes more difficult when complete workflows are networked with each other, resulting in highly complex automation solutions in the form of an R&D plant. The final escalation stage of laboratory automation is the linking of locally different R&D laboratories, usually for globally operating companies.

A few years after the turn of the millennium, three automation suppliers emerged in the paint and coatings industry market, which have since implemented and continuously developed many systems. These are the Swiss company Chemspeed Technologies, the German company Füll Lab Automation (which emerged from the former Bosch Automation) and the English company Labman Automation.

Detlef Gysau: Lab Automation and Digitalisation in Coatings

3.1 Simple automation solutions

Automation in the laboratory began decades ago. A classic example of automation is the autosampler of analytical instruments. Examples include gas chromatography, liquid chromatography, atomic absorption spectroscopy, ion chromatography and particle sizing, to name but a few. This simple repetitive task consists of transferring small vials and vials from the multiple sample holder to the starting point of an analytical process. This allows the analyzer to work around the clock and the lab technician to focus on the often time-consuming and manual sample preparation, see Figure 3.1.

Figure 3.1: Autosampler for analytical instrument
Source: SKW Stickstoffwerke Piesteritz

Automated pipetting is also used in the field of analytics and microbiology. Pipetting robots take over the monotonous, repetitive work and replace tedious manual pipetting, see Figure 3.2. This significantly increases productivity and reproducibility. The state of the art offers electronic pipettes from a few up to 384 channels, which enable the filling of 96- or 384-well microtiter plates. In synthetic biology, for example, efficient sample preparation for NGS (next generation sequencing) and suitable, reliable liquid handling systems are important for fast turnaround times and higher multiplex sequencing. With a suitable setup, one laboratory employee can realize more than 1500 sample workflows in 24 hours with a liquid handling platform [1].

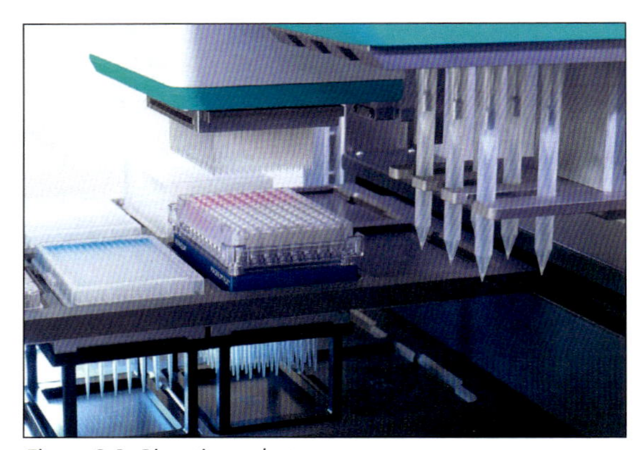

Figure 3.2: Pipetting robot *Source: SPT Labtech*

When handling substrates such as painted sheet metal, glass plates, plastic plates, wooden panels and wafers, robot systems simplify sample transfer and inventory management. Industrial robots or, increasingly, cobots can be used for this purpose. A collaborative robot, or cobot for short, is an industrial robot that works together with humans and is not separated from them by protective devices during the production process. The first cobots were developed in 1996 by *James Edward Colgate* and *Michael A. Peshkin*, two professors at Northwestern University in the US, and patented in 1997 [2]. The special feature of collab-

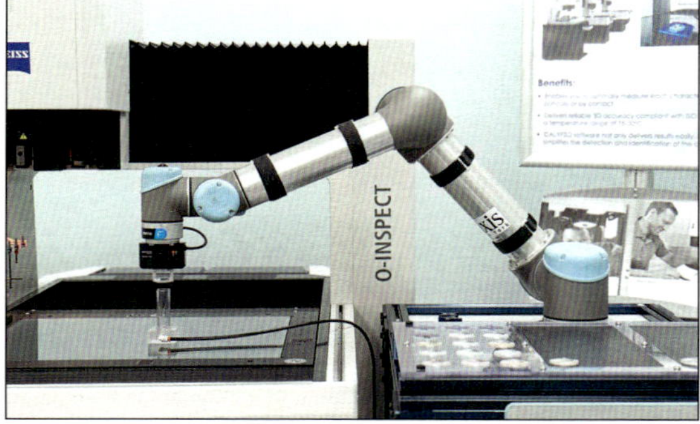

Figure 3.3: Pick and place Cobot for feeding Zeiss O-inspect
Source: Quality Magazine, BNP Media

orative robots is that they can work in close proximity to humans and together with them. This presupposes that the robots cannot cause injury to humans. Fences and other protective devices are then no longer necessary, as the robots have their own sensors that prevent injuries to human employees. The robots switch off automatically if they come into contact with obstacles, see Figure 3.3.

Robots can be used to efficiently load and unload test equipment. Frequently monotonous and constantly recurring processes, for example for long-term tests when loading and unloading painted substrates in rapid weathering devices or containers for testing storage stability to assess phase separation and viscosity, are actual logistics processes. Replacing these routine tasks relieves the scientific laboratory staff and frees up resources for much more demanding and changing tasks.

In larger paint and varnish companies, thousands of painted substrates are exposed to high-energy radiation combined with different climatic conditions (temperature and humidity) in accelerated weathering devices every year. Rapid weathering devices from companies such as Q-Lab (QUV for accelerated weathering testing) and Atlas (Weather-Ometer with xenon lamps) are widely used. In order to track their behaviour over the weathering period, the samples are removed from the test devices in predefined weathering cycles, prepared for measurement of the colorimetry and returned to the weathering test devices after the measurement. Depending on the requirements, the test can run from several hundred hours to several thousand hours. As a result, the weathering panels are loaded and unloaded several times until their cycles are finished. The use of autonomous mobile robots (AMR) even enables transportation from the conditioning cabinets or rapid weathering panels to the measurement automation.

The same applies to testing the storage stability of wet samples, which are stored under different conditions and tested for possible changes after defined cycles. Classically, phase separation in the form of syneresis or sedimentation formation and the change in viscosity are tested and assessed qualitatively and quantitatively depending on the requirements. This also requires a great deal of manual effort for the simple logistics and measurement tasks. The handling of wet samples can be reduced to simple and efficient manual transfers of trays from the conditioning cabinets to the automated storage system. From then on, the automation takes over the handling

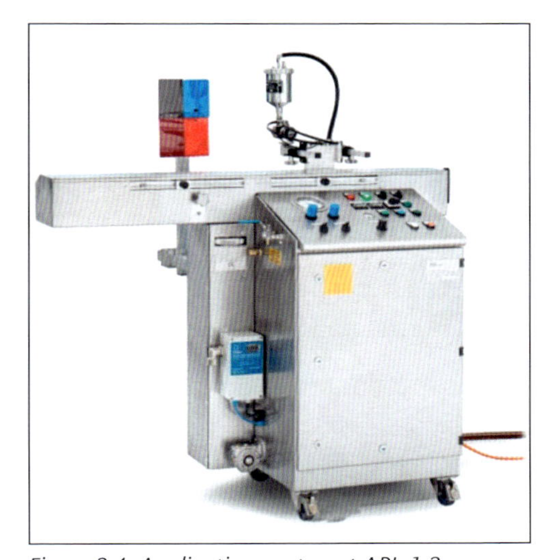

Figure 3.4: Applications automat APL 1.2
Source: Oerter GmbH & Co. KG

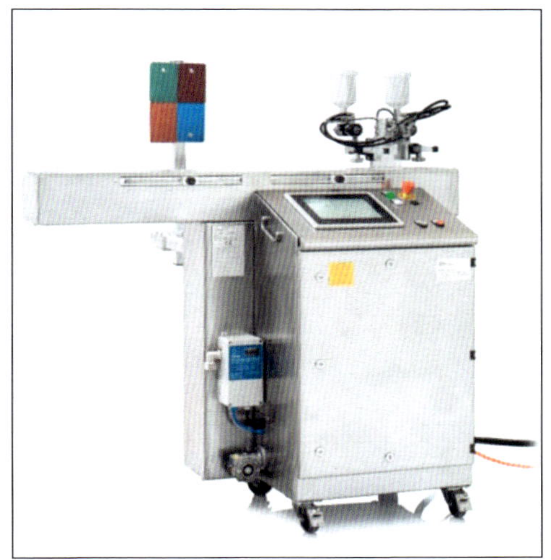

Figure 3.5: Applications automat APL 2.3
Source: Oerter GmbH & Co. KG

of the individual samples for the preparation and execution of the measurement, including sample recognition and data management.

Another form of simple automation, especially from the point of view of standardization, are spray robots for the uniform execution of spraying processes for test panels. These can be purchased pre-assembled for various solutions and configurations. However, the focus is often on standardizing the spraying process. These automation solutions still require human assistance, as they only represent partial automation. The manual tasks include filling, emptying and cleaning the spray gun, as well as assembling the test panels to be painted, see Figure 3.4 and Figure 3.5.

Another simplified solution for more individual spraying solutions are self-configurable cobots, some of which are built by paint and varnish manufacturers themselves. There are examples of this in the raw materials industry, but also in the paint manufacturing sector. Adler in Austria, for example, offers its customers the optimization and use of its products with a cobot coating, see Figure 3.6. It is difficult to achieve a uniform wet film thickness by manual application. If windows are sprayed manually, the paint consumption can be between 15 % and 30 % higher compared to a painting robot. A painting robot guarantees an even paint application and a constant speed. The constant distance between the spray guns and the product optimizes paint consumption and ensures a uniform coating. Painting robots can be used for water-based and solvent-based products. Optimum robot settings reduce paint consumption, minimize downtimes and thus optimize production. A painting robot is particularly suitable for large-format 3D parts such as doors

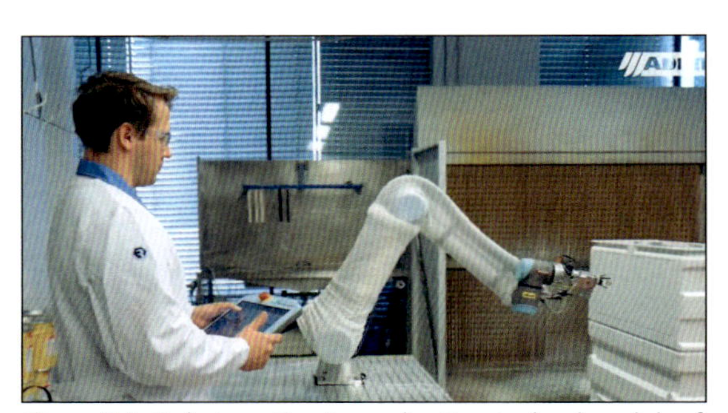

Figure 3.6: Cobot coating in application technology lab of Adler Source: ADLER-Werk Lackfabrik Johann Berghofer GmbH & Co KG

Figure 3.7: High throughput experimentation with individual automation islands

Source: Clariant Innovation Centre (CIC)

or window frames and enables you to produce components for the wood or furniture manufacturing process with a consistently high surface quality.

In contrast, Clariant has many different stand-alone solutions at its Innovation Centre in Frankfurt, which can operate independently of each other, but can also be networked to a limited extent depending on the configuration, see Figure 3.7. These can be more complex synthesis reactions for polymers, but also tests of physical-chemical, mechanical or rheological properties.

3.2 Stand-alone automation solutions

In addition to the automation of analytical measuring devices for sample preparation and sample transfer processes, an automated stand-alone solution for tests or application processes that need to be repeated in large numbers is often the entry point into laboratory automation. Stand-alone solutions include far more than just the execution of a process as previously described in Chapter 3.1. This means that the automation also includes all peripheral processes, which are carried out independently once the automation system has been loaded with a large number of samples and substrates. The financial outlay is also kept to a manageable level. However, the fundamental question for the investment also applies here. Where is the biggest bottleneck in my laboratory processes? What are "dull" routine tasks in the laboratory for which I am wasting valuable and qualified resources? Which processes have major differences due to the human factor and can be standardized with increased quality through automation? There are many examples, some of which are presented in the following. Depending on the manufacturer, different concepts are offered with their advantages and disadvantages. Stand-alone solutions very often also consist of partially automated concepts, i.e. manual intervention in the system may be necessary depending on the process step.

3.2.1 Paint formulation solutions

Universal solutions for various markets are offered on the market. In addition to applications for paints and coatings, related industries such as the adhesives and sealants industry, home care, beauty care and batteries, to name but a few, are also being served. In principle, this is a possible approach, as all industries involve the dosing of liquid and solid raw materials and subsequent dispersion. The main differences with liquid raw materials lie in their viscosity and with solid raw materials in their flow behaviour and particle size with regard to the smallest, precisely metered quantity by weight.

A stand-alone solution for the formulation of paints and coatings represents the "Labman Formulator", a modular formulation system with comprehensive and flexible dispensing capabilities, see Figure 3.8. Individual dispensing stations can accommodate a range of dispensers, from liquids and pastes to pow-

Figure 3.8: "Labman's Formulator" with raw material containers in the back *Source: Labman Automation*

ders. The system can be set up to contain more or fewer powder or liquid dispensers. Each dispenser type can be configured according to the materials and dosing tolerances contained. The system allows periodic agitation of viscous liquids. The dispensing pressure can be pulsed for more precise dispensing and a peristaltic pump enables regular agitation of viscous liquids. The dispensing hose is squeezed via rollers to dispense the liquid. Precise dosing of liquids is achieved by means of syringes and a dosing needle, which can be removed from a reservoir. Once all the raw materials have been added to the target container, it is closed again and placed in a dual asymmetric centrifuge (DAC). The DAC mixes the contents of the container at a configurable speed and duration. After mixing, the target container is placed back in the rack. The process with DAC is suitable for screening formulations, but is not necessarily scalable to production scale, as there is no mixing during the entire dosing process. Depending on the type of DAC used, it can also happen that not all solid surfaces are wetted with liquids before dispersion takes place, especially with high solid contents. This is manifested by possible lump formation in the lower cup area. The throughput of the system varies depending on the number of recipe components and batch size.

Chemspeed Technologies offers a similar stand-alone solution with its "Formax" in two size variants. The "Formax" for paints and coatings is a modular robot platform that enables automated, high-quality formulations for paints and coatings, see Figure 3.9 and Figure 3.10. Designed with flexibility in mind, it allows new tools, racks, vessels, mixers, application and testing tools to be modified and/or added at any time and currently offers a choice of up to 70 tool functions. One major difference, however, is the mixer used. This is more similar to mixer designs used in production, with agitators in the base area and rotating blades on the container wall, and also allows the addition of liquid and solid recipe ingredients during mixing. Depending on the configuration, the system offers space for between 3 and 24 dispersing stations (formulation reactors) and a size of 100 or 1000 ml. In contrast to the use of DAC mixers, the temperature of each formulation reactor can be precisely controlled with this concept, see Figure 3.10. The formulation reactors must be emptied and cleaned manually after production. The automation solutions can also be

Figure 3.9: Chemspeed's "Swing Formax"
Source: Chemspeed Technologies

Figure 3.10: Chemspeed's "Flex Formax"
Source: Chemspeed Technologies

extended with extensions or transfer stations, for example for the characterization of physical-chemical properties or application processes.

With its stand-alone solution, Füll Lab Automation offers simple, fully automatic preparation of formulations such as colours, mixtures, creams, etc. It is suitable for all types of liquid and paste-like formulations requiring only a small footprint of 3 m^2. Dosing is gravimetrically controlled for all types of liquids and powders as well as for volumetric dosing of liquids from 18 raw material positions on a carousel, which can flexibly be used for liquids in BLS syringes, double pressure-time-(pt)-valves for liquids in pressurized containers (2 raw materials at each position) or powder containers. The system does not require cleaning of the 50 available formulation containers as these are consumables (BLS-cylinders). The "Compact Lab Station" can be expanded with additional formulation, application and characterization stations, e.g. pH and viscosity measurement and adjustment, see Figure 3.11.

3.2.2 Paint application solutions

Once the paint and lacquer samples have been produced, the next process steps follow. These workflow steps can be carried out in isolation with application platforms, regardless of whether it is a doctor blade application or spraying process. One of the pioneers of an automated spraying system was Bosch Automation, which was merged into Füll Lab Automation in March 2021. Bosch invented "BLS syringe" Technology back in 2003. The "BLS syringe" is used to introduce the coating material into the spray application, see Figure 3.12. This syringe is comparable to a cartridge system that can hold and store material and supports direct spraying without decanting the coating material. The dosing tip is inserted into the spray head in such a way that there is no direct contact between the liquid coating material and the spray nozzle. Therefore, no cleaning is required between the individual, horizontal spraying processes. The "BLS Spray Application" is based on an award-winning technology that won the European Coating Award in 2005 and has since been supplied to numerous paint shops. Over the years, this technology has been further optimized and supplemented with additional features. It is available in different sizes with a throughput of up to 100 or a maximum of up to 400 coatings in different substrate sizes and can also be integrated as part of a fully automated R&D system for paints and coatings, see Figure 3.13.

Figure 3.11: The "Compact Lab Station"

Figure 3.12: "BLS Spray Application"

Source: Füll Lab Automation

Chemspeed Technologies has also been offering stand-alone spray automation for more than a decade. This is also available in different versions. Two variants are available in different sizes as "Flex Airborne Spray" and "Flex Airborne Spray L". This enables the application of different formats and numbers of substrates, see Figure 3.14. Furthermore, the systems can supply the spray applications in horizontal and vertical versions on request. Similar to Füll Lab, Chemspeed has also patented a development that allows clean-free and contamination-free application. Consumables are also used for this, but conventional syringes which are offered by various manufacturers. After aspiration of the coating material, these are guided precisely in front of the spray head and sprayed onto the substrates using atomizing air and a horn barrel. The software used also allows the spraying of multi-layer structures, intermediate flash-off or wedge-shaped layer thickness structures.

Depending on the configuration, different numbers of wet samples, substrates, syringes and sizes are available to the user on the system. The duration of unattended operation of the system depends on this. Of course, the systems are also available in an explosion-proof version, similar to Füll Lab Automation. Chemspeed Technologies also offers integration into fully automated systems for R&D for paints and coatings, see Figure 3.15 and Figure 3.16.

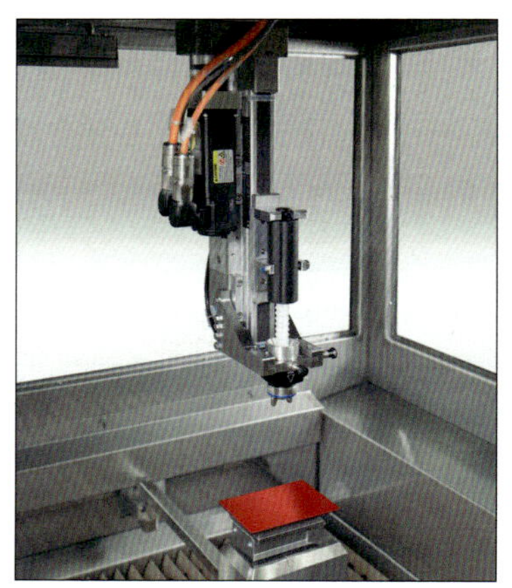

Figure 3.13: "BLS Syringe" with horizontal substrate positioning
Source: Füll Lab Automation

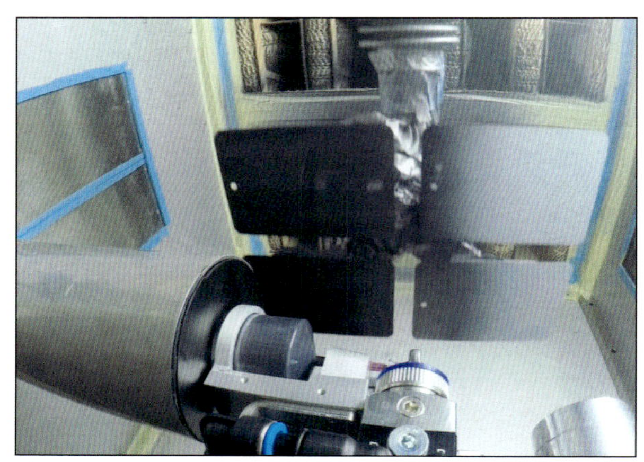

Figure 3.14: Syringe with coating material in front of spray head in 4-panel spray cabinet setup
Source: Chemspeed Technologies

Figure 3.15: "Flex Spray Airborne"
Source: Chemspeed Technologies

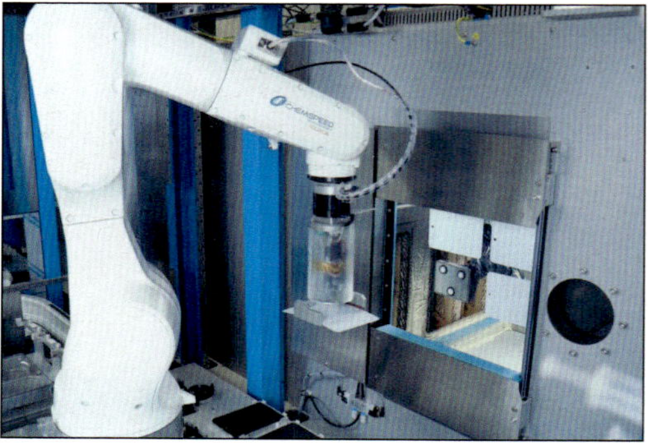

Figure 3.16: Loading of four substrates into spray cabinet
Source: Chemspeed Technologies

In addition to spray application for paints, drawdown application is used at least as frequently for architectural paints. Chemspeed Technologies offers a wide variety of drawdown applications to the market. The portfolio includes classic drawdown bars, but also drawdown bars with one or two chambers, all available in different gap heights, see Figure 3.17 to Figure 3.19. This is ideal for parallel drawdown of a sample in comparison to a reference sample, for example in quality control. For filling, the wet samples are poured directly into the chambers. Alternatively, the wet sample can be aspirated from the cup and dispensed onto the substrate using conventional syringes. Various materials such as cardboard, foils, metal sheets, glass or wooden panels can be used as substrates. The application is carried out using either a linear axis or a multi-axis robot. The applied substrates are then transferred to storage or transferred to heating plates or curing ovens for optional drying processes. After application, the drawdown bars are automatically transferred to the cleaning station and are then available for the next application, while the next coating films are applied in the meantime with the second drawdown bar set.

The Labman drawdown system dispenses sample from a pre-filled syringe onto the desired surface, consisting of an XYZ gantry and stacker axis to receive drawn down panels. The process starts with the system picking a syringe, reading the barcode and de-capping it. For slide drawdown, a die is placed on the syringe tip. The system dispenses a measured amount of sample

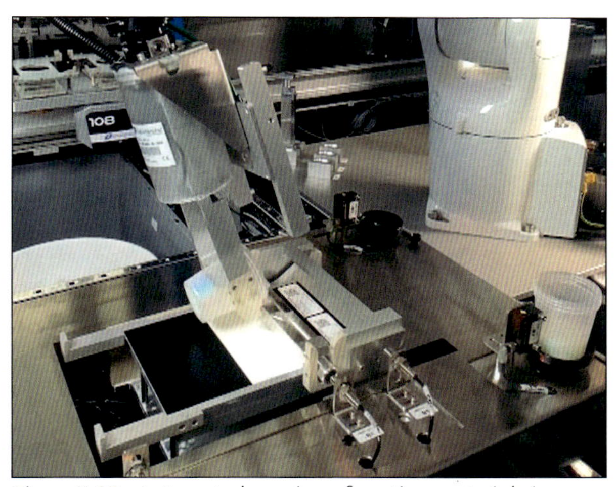

Figure 3.17: Automated pouring of coating materials in double frame bar Source: Chemspeed Technologies

Figure 3.18: Automated draw down of test and reference sample in double frame bar Source: Chemspeed Technologies

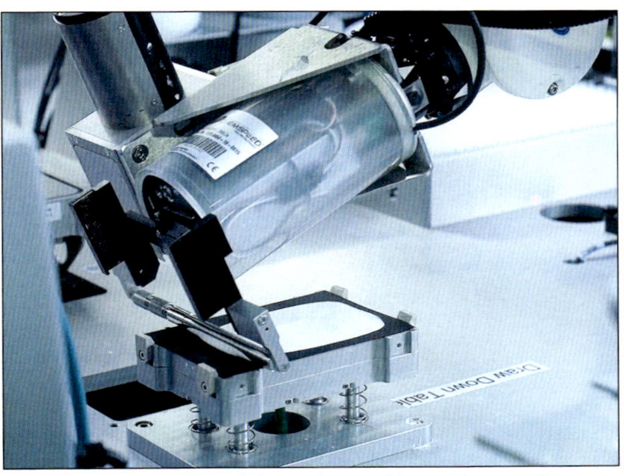

Figure 3.19: Automated draw down with gap bar performed by multi-axis robot Source: Chemspeed Technologies

and draws downs using the die onto 20 glass slides secured in a universal slide rack. For panel drawdown, a panel is placed on the vacuum bed where the syringe dispenses a measured amount of sample. According to the requested film thickness the robot picks up the corresponding draw-down bar, which is then moved across the panel creating a thin film of sample. The panel is pushed into the stacker and the process is repeated. Unfortunately, Labman offers only a semi-automated cleaning system requiring still some manual action, see Figure 3.20 und 3.21.

3.2.3 Paint characterization solutions

The testing of different wet paint properties, but also the characterization of physical-chemical, mechanical or optical properties are available as stand-alone systems on the market. For certain tests, such as the flow behaviour of paints and coatings, there are several suppliers on the market with different configurations. In some cases, viscometers and rheometers from different manufacturers such as Brookfield, TA Instruments and Anton Paar are integrated into the automation

Figure 3.20: Multi-functional system for producing moulds and thin film drawdowns on steel panels and glass slides *Source: Labman Automation*

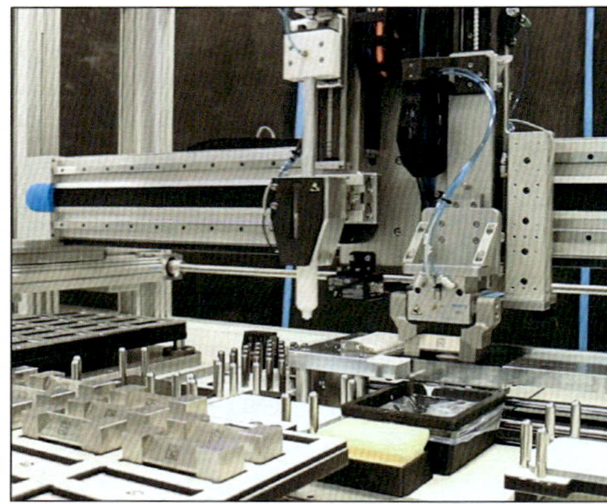

Figure 3.21: Detail view of draw down table and bar storage *Source: Labman Automation*

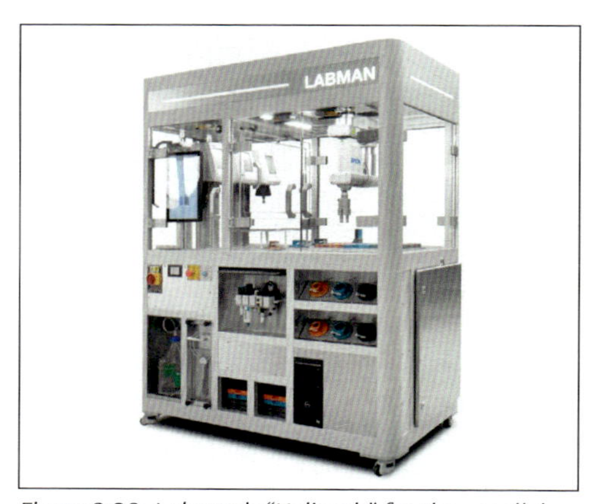

Figure 3.22: Labman's "Helipath" for the parallel measurement of viscosity *Source: Labman Automation*

Figure 3.23: Detail view of "Helipath" setup with three Brookfield viscosimeter *Source: Labman Automation*

systems. However, there are also suppliers such as Anton Paar who offer their own automation solutions for their testing devices on the market.
Labman Automation offers an elegant solution for testing viscosity. The system contains 3 viscometers that take measurements of 3 different size sample jars. The jars are loaded into the system and their locations are entered on the user interface (UI). Once the user presses start the robot arm picks the sample, scans the barcode, and places it into one of the appropriate shuttles. The shuttle then moves with the wet sample under the viscometer where a measurement is taken. Once the measurement is finished the shuttle moves back to the robot arm and the sample placed back to the rack. The viscometer lowers into a wash bath where the geometry is washed and dried. The viscometers can be setup to use either all the same jar type or a combination of different types. Shuttles on magnets and quick release airlines and racks are colour coded as to provide clear feedback on which jar size each viscometer is setup to use, see Figure 3.22 and 3.23.

Chemspeed's portfolio also includes a stand-alone system for measuring viscosity, optionally with additional dosing of raw materials and their homogenization. The proprietary overhead gravimetric dispensing technology allows the automatically dispense virtually of any product from low to high viscosity liquids, pastes, even melted waxes and subsequently measure the viscosity. Furthermore, the concept allows to measure viscosity virtually everywhere, e.g. formulation vessel, sample vials. The robot with its tool exchange design picks up a Brookfield viscosimeter and moves it to the sample place for direct measuring without sample transfer, see Figure 3.24. Alternatively, Chemspeed is also integrating the "Rheolab QC" from Anton Paar. In addition, it allows flexible sample conditioning for repeatable high-quality measurements. Features of the system include the automated opening/closing of sample containers, temperature-controlled sample while processing, range of viscosity measurement, type of viscometry and measuring geometries as well as cleaning of measuring geometries, see Figure 3.25.

In the industry, rheometers are used to obtain more detailed information about the flow behaviour of paints and varnishes over a wide range of shear rates. There are several suppliers with different approaches in this area as well. Anton Paar rheometers are widely used and are now also available in automated versions. Anton Paar pursues two different concepts. The "HTR 3000" is

Figure 3.24: Overhead Brookfield viscometer is moving by the robot to the sample for measurement
Source: Chemspeed Technologies

Figure 3.25: Overhead Rheolab QC viscometer is moving by the robot to the cleaning station
Source: Chemspeed Technologies

equipped with the "Bob Cup" system for measurement and, depending on the configuration, holds a maximum of 54 pre-filled cups that are sealed with temporary lids during storage, see Figure 3.26. This prevents volatile components of the wet sample from evaporating. After the lid is removed, the cup is moved by robot to the classic tabletop rheometer 102e or 302e and positioned before the upper measuring geometry (bob) is immersed. There is an integrated cleaning unit for the bob. The cups must be emptied and cleaned outside the automation. Anton Paar offers the "HTR 7000" for measuring paints and varnishes using plate-plate or plate-cone geometries. This includes a dispensable sample system for liquid samples with up to 1000 mPas that are suitable for exact dosing. Furthermore, there is a system for concentric cylinder measurements for low-viscosity samples and a system for pasty or gel-like samples that are not suitable for exact dosing. Optionally, the pH of the wet samples can also be measured.

Chemspeed's "Flex Rheo" allows fully automated rheology measurement for various workflows, using "Discovery HR" rheometers from TA Instruments, or other rheometers, for "Press Start and Walk Away" operation, see Figure 3.27 and 3.28. With optional extras, including robotic pH measurement tool, screw capper and extended storage capacity for different sample sizes, this Chemspeed configurable solution transforms the traditional rheology analysis process. The system allows to have three different measuring geometries – plate-plate, plate-cone and bob-cup – at the same time on the system. Included is also an active washing for all measuring geometries. Furthermore, it comes with an anti-vibration setup, reproducible sample dispensing, minimized cleaning liquid consumption as well as with easy export of data and fast integration into lab environment.

Chemspeed has developed various techniques allowing the aspiration and dispensing directly to the measuring geometries. The dispensing of very high viscous materials is also possible by using the paste dispenser. For this, the lower measuring geometry is positioned under the paster dispenser, which is then ejecting the paste onto the centre of the lower geometry. Afterwards the robot moves the lower plate to the rheometer to prepare the start procedure. The trimming tool automatically scrapes the overdosed sample material away from to upper geometry ensuring an accurate and reproducible measurement. Like in manual operation, the system can be equipped with a solvent trap and Peltier element for precise temperature control, see Figure 3.29 to 3.31.

3.3 Connected automation solutions

In addition to the increasing popularity of fully automated R&D facilities, combined with frontrunners from large companies, there are of course more profound reasons such as streamlining

Figure 3.26: Benchtop Rheometer HTR 3000 with MCR 102e/ MCR 302e *Source: Anton Paar*

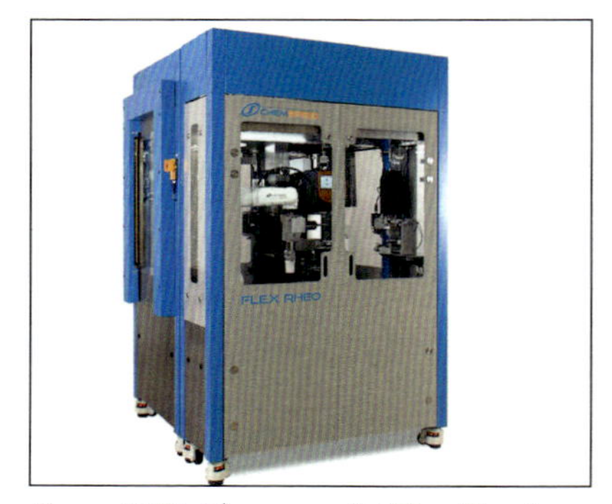

Figure 3.27: Chemspeed's "Flex Rheo" *Source: Chemspeed Technologies*

processes, reducing costs and at the same time increasing productivity and shortening time to market. At the same time, it eliminates capacity bottlenecks and the problem of recruiting qualified personnel. Overall, fully automated R&D laboratories are a strategic investment for companies [3] that want to maximize their productivity, quality and safety while preparing for the future of scientific and technological progress. However, it is also essential that sufficient space is available for a fully automated R&D facility. This aspect is often underestimated, which is why many future users also invest in a new property. This has resulted in quite a few so-called innovation centres. Smaller, fully automated systems require 40 m^2 of space, including the necessary free space around them, in order to be able to carry out daily tasks such as loading consumables and raw materials, as well as periodic maintenance work. The largest fully automated R&D plant is currently located at Altana's Byk-Chemie and takes up an area of more than 300 m^2.

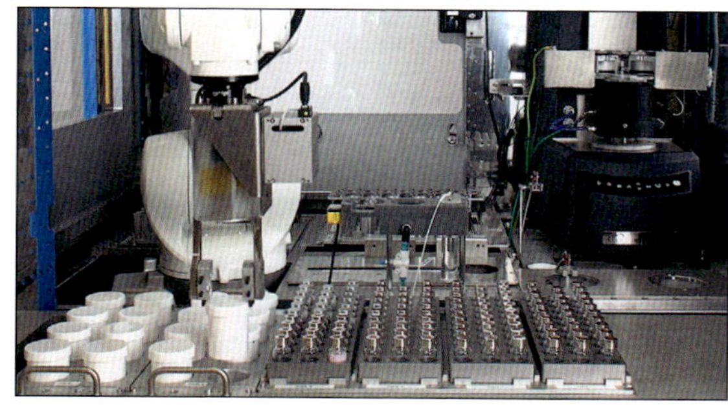

Figure 3.28: Partial overview of exemplary setup with TA "Discovery HR-202, screw capper, pH measurement & cleaning station *Source: Chemspeed Technologies*

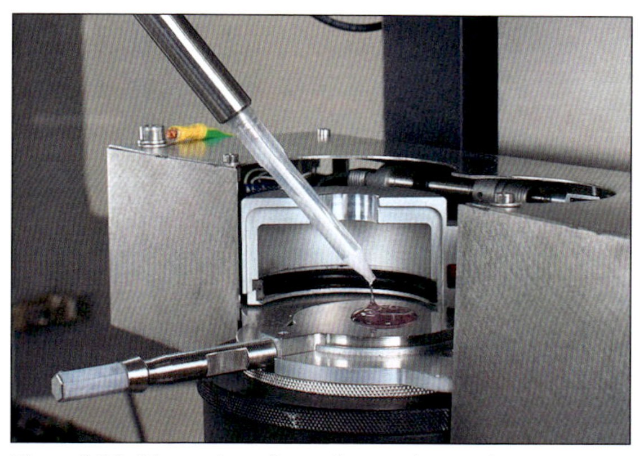

Figure 3.29: Dispensing of sample onto lower plate geometry *Source: Chemspeed Technologies*

Figure 3.30: Paste dispensing on the lower measuring geometry *Source: Chemspeed Technologies*

Figure 3.31: Trimming of high viscous paste after positioning of upper and lower plate-plate geometry *Source: Chemspeed Technologies*

Large international companies in the paint and coatings industry began their digital transformation back in the 2000s and invested in automated R&D facilities. However, the range of products on offer back then is not comparable to today, some 20 years later. The range of hardware has been constantly expanded and there are much more powerful systems, particularly in the area of software. Of course, advances in robotics have also contributed a great deal to new developments, which is why large, room-filling automation systems are significantly more powerful and faster today. Not all companies use their investments to promote themselves to their customers. However, this usually has a positive effect on shareholders, as it is a sign of progress, performance and sustainability in equal measure. The following companies or universities in the paint and coatings industry, listed in alphabetical order based on public media reports or referencing, have been investing in laboratory automation for many years:

- Adler Lacke [4]
- AkzoNobel Coatings [5]
- BASF [6]
- Byk-Chemie [7, 8]
- Clariant [9]
- Dow Chemicals [10]
- Evonik [11]
- Innospec [12]
- Merck [13]
- Omya [14]
- PPG [15]
- Sakata Inx [12]
- Tronox [12]
- Universities of Colorado Boulder [12], Ghent [12], Krefeld [16], London [12]
- Venator [12]
- VLCI [17]
- and many more not like to be listed.

Chemspeed's “Flexshuttle” is a solution that fundamentally changes workflows in product development and quality control of paints and coatings. This solution can be easily expanded at virtually any

Figure 3.32: Chemspeed's “Flexshuttle” Paints & Coatings at Omya in Switzerland

Source: Chemspeed Technologies

Figure 3.33: Chemspeed's “Flexshuttle” Paints & Coatings at Evonik in Essen

Source: Chemspeed Technologies

time (higher throughput, changed specifications, functions, workflows, etc.). The modular design enables a significant increase in efficiency combined with high quality and maximum ease of use. The automated laboratory can be individually modelled, see Figure 3.32. The operator navigates shuttles on a rail system to transport samples and substrates. An undefined number of work processes can therefore be carried out and controlled in parallel or synchronized. Liquids, pastes and solids are dispensed gravimetrically from small to large quantities with high accuracy. Methods such as doctor blade application, airborne spraying, spin coating or roller coating can be selected and combined for the application. Various options are available for drying, including simple drying at room temperature, forced drying on heating plates, curing ovens and radiation curing (UV/IR). The choice of test methods seems endless, as more than 200 devices and instruments from well-known manufacturers are now integrated and controlled by the Chemspeed workflow management software. Examples of liquid characterization include pH measurement and adjustment, density, viscosity, rheology, particle size measurement, flow behaviour, defoaming, etc. Applied coatings can be measured using various methods, e.g. colour measurement, gloss, surface texture, surface tension, turbidity, hardness, abrasion resistance, stain resistance, chemical resistance, to name but a few.

The multitude of individually and synchronously operating modules typically includes metering modules for small and large quantities of liquids such as polymers and additives as well as solids such as pigments and mineral fillers. Throughput is increased many times over, as it is at other "Flexshuttle" users such as Adler Lacke, Byk-Chemie [7, 8], Evonik [11] and Omya [14]. The industrial companies report an increase in sample throughput of a factor of 5 to 80 compared to manual laboratory operation, depending on the complexity of the formulations and the operating time, in the best case 24/7 and also on weekends. As already mentioned, there are occasional space problems when integrating a large HTE plant into existing buildings. This is where the Chemspeed "Flexshuttle" concept offers advantages. Two adjacent rooms can easily be connected. Only two small "windows" in the wall are needed to pass the rails through. Thus, the two rooms can be connected to each other by the shuttle operation and still realize a large HTE plant. The whole thing can be enhanced by connecting the plant over two floors by means of a paternoster for the shuttles, see Figure 3.33 and 3.34.

Figure 3.34: Chemspeed's "Flexshuttle" Paints & Coatings at Byk-Chemie in Wesel Source: Chemspeed Technologies

Figure 3.35: Labman's paint formulation system Source: Labman Automation

The "Labman" advanced formulation system has been designed in order to streamline the development of new products and ingredients. The system allows the scientists to rapidly build formulations consisting of many steps in any combination required. Formulations can consist of any combination of liquid dispenses, powder additions, water dispenses, mixing using multiple topologies, pH measurement and automated adjustment, viscosity measurement and automated adjustment, delays, timers and temperature control. This system uses new and leading technologies to deliver a robust platform that can be used in very flexible ways by its operators, see Figure 3.35. Its flexibility allows it to run in ways that best optimise the samples being formulated or tested. The system can be used to formulate new raw ingredients that can be used as raw materials additions, to make full or partial formulations, or to measure existing samples for categorisation and calibration.

Figure 3.36: Transfer of functions, tools, samples or substrates by trolleys Source: Labman Automation

The Labman concept is based on combined stand-alone solutions that use trolleys to transfer functions, tools, samples or substrates between the modules. For example, substrates applied in the Formulation Module are stored in a storage system mounted on a trolley. After filling, the trolley is rolled out of the Formulation Module into the next stand-alone module, for example to characterize optical or mechanical properties, and fixed in position. The substrates are then unloaded and subjected to tests.

Figure 3.37: Two multi-axis robots moving on the linear axis to serve different functions Source: Labman Automation

Alternatively, there are also large system installations that are operated with one or two multi-axis robots on linear axes. The linear axis with a few multi-axis robots is obviously a cheaper alternative to a rail system with switches and shuttles, but the consequence is that adapting the system layout depending on the room design is more difficult or allows less flexibility. Another disadvantage, probably much more decisive, is a reduced throughput because fewer multi-axis robots work in parallel compared to Chemspeed's "Flexshuttle" concept. Labman compensates for this with

Figure 3.38: "Integrated Lab System" of Füll Lab Automation Source: Füll Lab Automation

additional "hidden" underground linear axes for transporting samples. Furthermore, the size of the storage system for the number of raw materials and raw material quantities as well as substrates is also decisive for the duration of unattended operation of the system, see Figure 3.36 and 3.37.

A newly introduced high-throughput experimentation system of Labman at AkzoNobel in Shanghai [5], the third HTE within AkzoNobel by Labman, integrates four core functions of raw material performance: screening, product performance testing, formula model design and colour formula database construction, as well as two modules of wet film preparation: optical and durability testing, which can fully meet the core testing needs of paint and coatings. It has powerful data processing capabilities, continuous test operation and self-check functions, which can not only improve research and development efficiency and test accuracy, but also help to further reduce experimental waste. Using this device enables AkzoNobel to reduce paint waste discharges by at least 50 % during new product development compared to similar tests. Exactly this is nowadays also an important driver for the investment of high throughput systems in laboratories as typically the sample size from manual formulation of 1 l is reduced to approximately 100 ml, as this is most of the time sufficient for the execution of tests and applications. However, users sometimes criticize the higher consumption of consumables in automation systems, which is contra dictionary to the raw material reduction of the systems.

Füll Lab's "Integrated Laboratory Station" (ILS) is a highly flexible system for automating research or quality control processes, see Figure 3.38. The "ILS" can be fully customized to requirements. Dosing raw materials, mixing, grinding and dispersing, adjusting properties such as pH and viscosity, characterizing the liquid, pasty or creamy formulation, using a formulation as a raw material, checking stability, spraying, draw downing or applying the formulation in other ways. All these functionalities are available as standard modules. But there is more: standard modules for the characterization of samples created by applying the formulation to a substrate, including destructive tests. The fully automated ILS is particularly suitable for large sample volumes. To ensure a high throughput, an ILS is equipped with one or more multi-axis robots, if required also with additional XYZ handling systems to further shorten the cycle time. According to the manufacturer, the efficiency of laboratory work is increased by a factor of approx. 2.5, which is significantly lower compared to the previously mentioned manufacturers. Nevertheless, development costs and waste disposal costs can be reduced by up to 60 %.

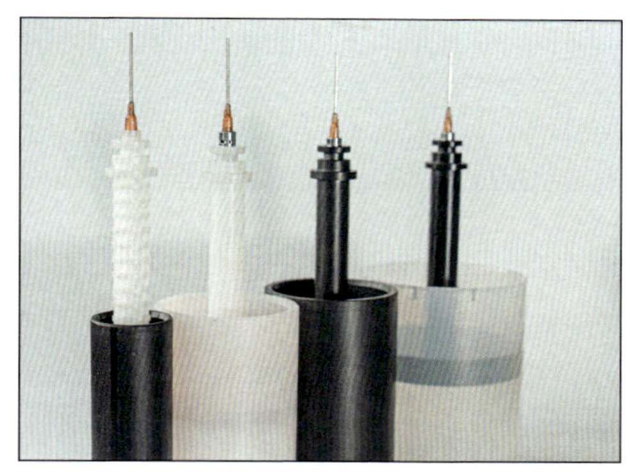

Figure 3.39: "BLS syringe" with dosing needle
Source: Füll Lab Automation

Figure 3.40: Loading of "BLS formulation syringe" into DAC
Source: Füll Lab Automation

The modules have their own controllers for complete decoupling of the workflow of the handling operations. The system allows completely flexible workflows for each sample, i.e. each sample can be different. The set-up also supports loops, for example: iterative powder dosing. Modules can be easily exchanged if requirements of the workflow changes as well as multiple ones be combined to increase storage capacity and througput. At the end of such a workflow is a formulated colour or coating. The basis for the "ILS" is the "BLS syringe", see Figure 3.39 and 3.40. The patented "BLS syringe" consists of a syringe cylinder and a piston with a dosing tip and is available in different sizes. The cylinder can be used as a formulation container and, after the piston has been inserted, allows the direct dosing of the prepared formulations. This means that formulations become new raw materials without the container having to be changed. In addition, the syringe can be used directly for application. The liquid formulation can be characterized directly in the syringe cylinder, i.e. there is no need to decant the sample. The syringe is a disposable item, thus eliminating the need to clean the formulation containers. A new container for each sample means that there is no risk of contamination. When used as a raw material container, the cylinder can be refilled and the syringe reused multiple times. However, based on the workflow requirements other containers, such as glass vials or standard lab cups with screw caps can as well be used. Besides the three before mentioned suppliers of HTE equipment for connected automation solutions, the offer for the coatings industry is weak. However, in other markets such as in life science, particularly the pharmaceutical industry, a few other players are present [3].

Figure 3.41: JAG MoMa 4 Mobile Manipulator AMR
Source: JAG

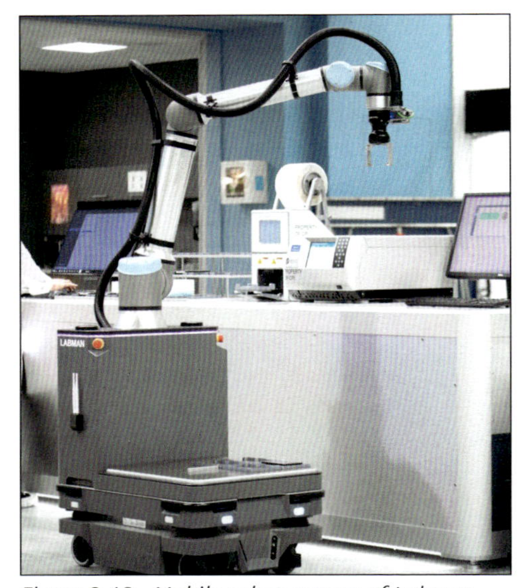

Figure 3.42: Mobile robot system of Labman
Source: Labman Automation

3.4 Connected automated laboratories

Connected automated laboratories, also known as "connected labs" or "smart labs," leverage a network of digital and automated tools, allowing for seamless communication, integration, and operation across different lab equipment, data sources, and processes. This could become valid for "isolated" labs at the same location, which initially are not connected to each other. Using autonomous mobile robots (AMR) could overcome the disconnection and trigger for higher speed and unattended operation of the isolated automation systems. Typical duties of AMR's are transfer jobs to ensure the continuous logistics between stand-alone systems and/or storages, see Figure 3.41 to 3.43. In an ideal world the AMR could even feed the automation systems with raw materials, substrates and consumables. In a larger scope they could also connect laboratories located in different parts of a plant. For example, delivering samples from production to an automated quality control laboratory. In other industries and applications AMR are indispensable, e.g. assembly lines in the automotive industry.

The larger scope of connected automated labs are in the sense of connecting laboratories based at different sites, countries and continents. Connected labs bring together data from various instruments and processes into a centralized platform. This centralization enables real-time access to comprehensive datasets, improving the ability to analyse results and make data-driven decisions. Thus, researchers can monitor and control experiments from anywhere via cloud-connected systems. This allows for flexibility in lab management, including real-time troubleshooting, progress checks, and remote intervention, if necessary. Also, connected labs make it easier for teams across different locations to collaborate on experiments, share data, and access lab resources in real-time. This is particularly valuable for organizations with geographically dispersed teams or partnerships with external research groups. By having connected systems that log and monitor every step of an experiment, it becomes easier to reproduce results and verify findings. This is essential for quality assurance, regulatory compliance, and knowledge transfer. By having connected systems that log and monitor every step of an experiment, it becomes easier to reproduce results and verify findings. This is essential for quality assurance, regulatory compliance, and knowledge transfer. Especially for quality control requirements, it contributes heavily to standardization across the different sites. A perfect example for this, was the announcement of Merck Surface Solutions for the quality control of their pigment productions on three different continents [13]. In addition, through connectivity, labs can optimize workflows by dynamically assigning tasks to available equipment or scheduling experiments based on available resources. This maximizes equipment utilization and reduces downtime. Finally, connected labs can monitor energy consumption, resource usage, and waste production in real-time. This data enables labs to make adjustments that improve sustainability by reducing excess resource use, minimizing waste, and enhancing energy efficiency, see Figure 3.44.

Figure 3.43: Autonomous mobile robot of Chemspeed *Source: Chemspeed Technologies*

Connected automated labs typically consist of a variety of integrated systems, including:

- **Internet of Things (IoT) Devices:** These sensors track environmental parameters (e.g., temperature, humidity), equipment status, and real-time experiment data.
- **Laboratory Information Management Systems (LIMS):** A digital platform

Figure 3.44: Global network
Source: Microsoft Networking Research Group

that manages, tracks, and integrates sample and experiment data, workflow management, and quality control.
- **Robotic and Automated Equipment:** Robots and automated systems conduct experiments, handle materials, and execute multi-step procedures.
- **AI and Machine Learning Platforms:** Tools for data analysis, anomaly detection, and predictive modelling based on accumulated lab data.
- **Cloud Computing and Data Storage:** Cloud-based systems enable data centralization, remote access, and secure data storage across different locations.

3.5 Literature

[1] PALYS, J., (seqWell Inc.); PAERSON, A.; WHITMORE, I.; REES, Huw, (SPT Labtech); "Hochdurchsatz in der Forschung", www.labo.de/dosier-und-vakuumtechnik/hochdurchsatz-in-der-in-der-ngs-probenvorbereitung.htm, published 7th May 2024

[2] Patent US5952796A: Cobots. Angemeldet am 28. Oktober 1997, veröffentlicht am 14. September 1999, Erfinder: James E. Colgate, Michael A. Peshkin.

[3] CHALLANER, C., "R&D Strategies in the Coatings Industry", www.paint.org/coatingstech-magazine/articles/industry-update-state-coatings-rd/, Coatings Tech of American Coatings Industry, issue August 2017

[4] bm-online.de, "Adler nimmt vollautomatische Laboranlage in Betrieb", www.bm-online.de/aktuelles/markt-branche/adler-nimmt-vollautomatische-laboranlage-in-betrieb/#, published 5th March 2013

[5] Labman Automation, "AkzoNobel implement third-of-its kind Labman robot", https://labmanautomation.com/customer-stories/akzonobel-implement-third-labman-robot, 2022

[6] "BASF and Bosch", https://doi.org/10.1108/prt.2006.12935eab.006, Pigment & Resin Technology, Vol. 35 No. 5.

[7] RECKTER, B., "Altana testet Lacke für die Industrie in weltweit einmaliger Hochdurchsatzanlage", www.vdi-nachrichten.com/technik/forschung/altana-testet-lacke-fuer-die-industrie-in-weltweit-einmaliger-hochdurchsatzanlage/, 15th February 2022

[8] Altana press release, "ALTANA sets new standards in Wesel: digital BYK laboratory is unique worldwide", www.altana.com/press-news/details/altana-sets-new-standards-in-wesel-digital-byk-laboratory-is-unique-worldwide.html, 3rd May 2022

[9] Clariant on YouTube, "High Throughput Screening (HTS/HTE) Method Explained", https://youtu.be/Q1GFtrm8hg4, 15th September 2017

[10] KUO, T.-C., et. al. , "High-Throughput Industrial Coatings Research at The Dow Chemical Company", https://pubs.acs.org/doi/10.1021/acscombsci.6b00056, Coatings Tech of American Coatings Industry, 21st July 2016, ACS Combinatorial ScienceVol 18/Issue 9

[11] RECKTER, B., "Lack vom Fließband", www.vdi-nachrichten.com/technik/produktion/lack-vom-fliessband/, 7th June 2018

[12] Labman Automation, "Paints and coatings", https://labmanautomation.com/industries/paints-coatings, labmanautomation.com, 2024

[13] KRIEGBAUM, E., Interview: "We see a need to catch up", European Coatings Journal, 13th December 2022

[14] Omya on YouTube, "Omya Lab | Flexshuttle automated formulation laboratory", www.youtube.com/watch?v=pJGQ3MzOVd8, 21st June 2024

[15] PPG press release, "PPG Opens Architectural Paints and Coatings Color Automation Laboratory in Milan", https://news.ppg.com/press-releases/press-release-details/2022/PPG-Opens-Architectural-Paints-and-Coatings-Color-Automation-Laboratory-in-Milan/default.aspx, 6th May 2022

[16] Press release University of Applied Sciences in Krefeld, "Hochschule Niederrhein macht mit Hochdurchsatzanlage Unternehmen fit für Industrie 4.0", www.hs-niederrhein.de/startseite/news/news-detailseite/hochschule-niederrhein-macht-mit-hochdurchsatzanlage-unternehmen-fit-fuer-industrie-4-0/, 30th November 2018

[17] MENNEN, S. M., et.al, The Evolution of High-Throughput Experimentation in Pharmaceutical Development and Perspectives on the Future, https://pubs.acs.org/doi/10.1021/acs.oprd.9b00140, Organic Process Research & Development 2019 23 (6), 1213-1242

4 Dispensing technologies

Dispensing technologies are essential for high-precision and repeatable dispensing in a variety of laboratory applications. As automation and connectivity improve, these technologies are increasingly integrated into lab workflows, supporting a range of needs from micro dispensing to large-scale material such as paints and coatings. Each technology has distinct benefits, making it important to choose the appropriate dispensing system based on specific research needs, sample type, and required dispensing precision. Precision becomes even more important in the case of high throughput experimentation as scale of experiments usually decreases compared with manual workflows. Considering the field of paints and coatings the size often decreases by factor 5 to 10, which means an increase of dispensing and weighing precision by factor 5 to 10. Typically dispensing of raw materials and intermediates in the paints and coatings industry are done gravimetric, especially in laboratory scale. The second most common way of dispensing is based on volumetric dispensing, but rather typical for production environment. Each of the dispensing systems has specific benefits and disadvantages. The mechanism for exercising the dispensing is based on different technologies such as micro dispensing pumps, syringe pumps, peristaltic pumps and pressure pumps. Both, pressure pumps and peristaltic pumps allow periodic stirring of viscous liquids. The dispense pressure can be pulsed for more precise dispenses for pressure pumps. In peristaltic pumps rollers squash the output tube to dispense fluid. Finally, syringe pumps dispense liquids from a reservoir via needle for higher accuracy dispenses.

4.1 Gravimetric dispensing systems

Gravimetric dispensing devices are crucial for precision and repeatability in various laboratory settings. They are reliable, versatile, and often easy to use, making them suitable for routine applications and high-throughput settings. Their ability to integrate with automated systems also makes them a foundational tool for modern laboratories, especially as labs shift towards higher automation and connectivity. Also, gravimetric dispensing in lab scale allows an easier upscaling into production.

The variety of gravimetric dosing technologies in laboratory automation is surprisingly diverse compared to manual weighing. Depending on the provider, different concepts are pursued and preferred. The origin can be debated but could well have reasons in patent-protected developments. In addition, there are also different concepts for raw material containers about consumables or reusability. Table 4.1 provides an overview without claiming to be exhaustive. Specific examples are therefore presented in the following chapters in alphabetical order of the suppliers, without any ranking.

4.2 Gravimetric liquid dispensing systems

The Chemspeed portfolio for gravimetric dispensing contains different sizes of cartridges and containers for liquids and powder materials. The choice of using an on-deck balance or overhead balance is considering the requested throughput requirements, also depending on if the dispensing tool serves a standalone unit or modules in the "Flexshuttle" system [1], Figure 4.1.

Table 4.1: Overview matrix of gravimetric dispensing systems and options

Dispensing from/with	**Weighing to**	**Type of balance**	**Balance**	**Materials**
Container	Single vessel	On-deck	Resolution	Viscous liquids
Container with stirring	Vessels in racks/trays	Overhead	Capacity	High viscous liquids
Container with heating	Vessel on shuttle			Solids (powders, granules, ...) with different particle size and shape
Cartridges	Formulation reactor			Pastes
Auge feeder/ vibration feeder				

Table 4.2: "Flex Liquid S" of Chemspeed Technologies

Key characteristics	**Description**
Dispensing concept	Direct dispensing into sample vessel on shuttle
Material	Viscous liquids such as additives, viscosity range: 1 to 4'000 mPas (for viscosities > 4'000 mPas, the heating option for dispensing units is required)
Size of feed container	240 ml
Storage capacity	Maximum 112
Recommended dispensing range	10 mg to 5 g
Balance resolution	1 mg
Balance	On-deck, automatic dynamic weighing

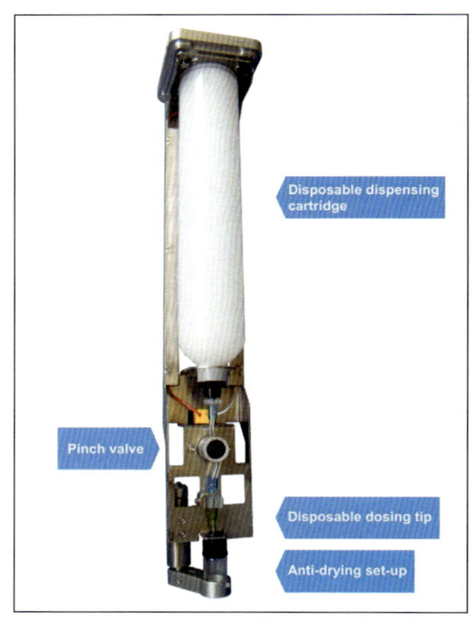

Figure 4.1: Disposable dispensing cartridge of "Liquid S" *Source: Chemspeed Technologies*

Figure 4.2: "Flex Liquids S" with storage on the left *Source: Chemspeed Technologies*

Table 4.3: Flex Liquid M of Chemspeed Technologies

Key characteristics	**Description**
Dispensing concept	Direct dispensing into sample vessel on shuttle
Material	Viscous liquids such as binders, viscosity range: 0.3 to 100'000 mPas
Size of feed container	1 l, 3 l, 5 l, optional with stirrer and/or heat jacket
Storage capacity	1 to 12 and 1 to 36
Recommended dispensing range	20 mg to 50 g
Balance resolution	1 mg
Balance	On-deck, automatic dynamic weighing

Table 4.4: "Flex Liquid L" of Chemspeed Technologies

Key characteristics	**Description**
Dispensing concept	Direct dispensing into sample vessel on shuttle
Material	Viscous liquids such as binders, hardeners, solvents viscosity range n/a
Size of feed container	19 l or 45 l, with stirrer
Storage capacity	1 to 10, optional expansible storage
Recommended dispensing range	20 mg to 100 g
Balance resolution	1 mg
Balance	On-deck, automatic dynamic weighing

The gravimetric dispensing system "Flex Liquid S", Figure 4.2 [2] is used as module in the "Flexshuttle" concept for small liquid volumes such as additives. The target vessel of different available sizes is transported via shuttle on the track system to the dispensing station. In parallel the overhead robot is mounting the proper cartridge from the storage hotel and moves it to the dispensing station. The shuttle is parked on the on-deck balance during dispensing. After one or multiple dispensing steps from different liquids, the shuttles with the sample vessel moves to the next workflow step. The "Flex Liquid S" can be used for the formulation of the mill base as well as for the let down of paints, because

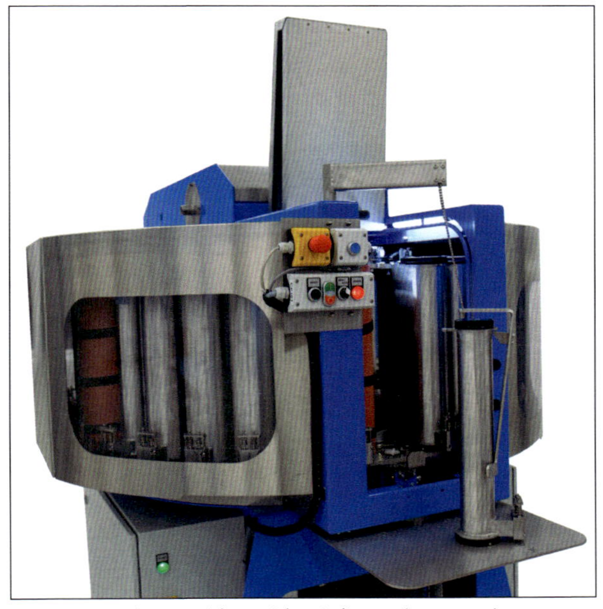

Figure 4.3: Flex Liquid M with 36 dispensing containers
Source: Chemspeed Technologies

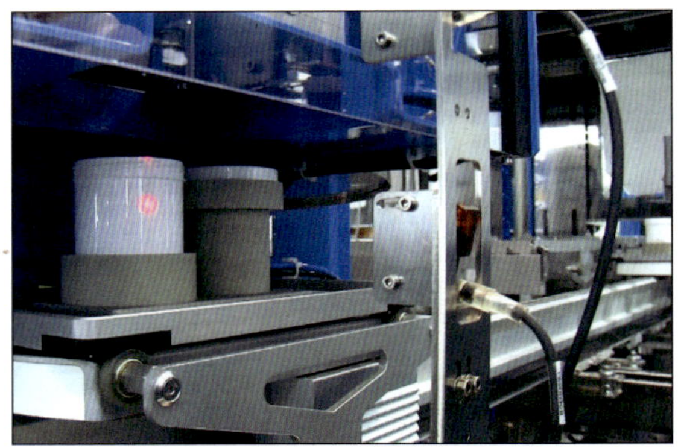

Figure 4.4: Sample vessel on shuttle parked below the dispensing station Source: Chemspeed Technologies

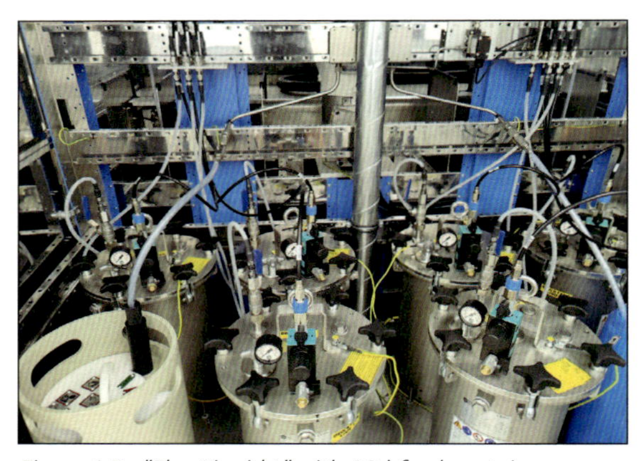

Figure 4.5: "Flex Liquid L" with 19 l feed containers connected via tubes to the dispensing heads
Source: Chemspeed Technologies

Figure 4.6: Dispensing station with battery of dispensing heads mountable to overhead robot
Source: Chemspeed Technologies

the shuttle with the sample vessel can take another loop on the "Flexshuttle" system to the "Liquid S" after the mixing respectively dispersing step.

The next larger gravimetric dispensing system of Chemspeed is called "Flex Liquid M", Figure 4.3 [3]. The system can be configurated manifold for dispensing containers of different sizes and further options such as stirrers and heat jacket. The heat jacket for keeps liquids heated during storage and allows an easier dispensing of higher viscous liquids.

The feed respectively dispensing containers can have volumes of 1 l, 3 l and 5 l and can be combined on the same dispensing system. The dispensing head is fixed to the bottom of each dispensing container. For the dispensing a specific container is moved by the revolving rack to the dispensing station above the shuttle with the sample vessel. The number of containers on the "Flex Liquid M" can be configurated by either 12 or 36, where the larger number will lead also required more space for the unit. Like the "Flex Liquid S", there can be one or multiple liquids dosed into the sample vessel. In addition, the sample vessel can return after mixing/dispersing to exercise the needed dispensing steps for the let down part, Figure 4.4.

The next larger dosing system from Chemspeed is called "Flex Liquid L", Figure 4.5 [4]. It is suitable for the direct dosing of larger, recurring liquid quantities. Gravimetric dosing is carried out dynamically and continuously via the built-in weighing module with a resolution of 1 mg. As with "Flex Liquid S" and "M", the sample vessel is located on the shuttle, which is parked on the on-deck scale under the dosing tool. (Viscous) liquids can be dispensed from a total of 10 pressure containers with a capacity

Table 4.5: "BLS" cartridge dispenser of Füll Lab Automation

Key characteristics	Description
Dispensing concept	Direct dispensing from the BLS syringe into formulation containers (e.g. BLS syringe)
Material	Low viscous liquids, to high viscous liquids; viscosity range 0.2 to 50000 mPas at room temperature, higher viscosities with heated syringe
Size of feed container	10 ml to 300 ml, other sizes available on request
Storage capacity	Unlimited amount of different raw materials possible. Quick exchange of raw material/BLS Syringes possible. A standard rack has 32 to 100 positions for BLS syringes. Several raw material racks possible depending on customized system design
Recommended dispensing range	10 mg to 100 g
Balance resolution	0.1 mg
Balance	On-deck, automatic dynamic weighing, volumetric dispensing possible

of 19 l each with stirrer. Each pressure container is connected to its own dispensing head via a material hose. The dosing head is coupled with an overhead robot and guided directly to the dispensing station, Figure 4.6. Pressure tanks of up to 45 litres and an additional storage unit can also be selected.

To prevent the dosing tip from drying out or hardening, there is an automated rinsing process. When changing raw materials, there is an integrated rinsing process and set-up for the raw material change. The "Flex Liquid L" is particularly suitable when very large quantities of the same liquid raw material are required more than 5 litres per day. This prevents refilling processes during a run without affecting the throughput.

The German company Füll Lab System prefers to use their "BLS syringe", Figure 4.8 [5] for the dispensing of liquids. The "BLS syringe" is in principle a cartridge with a plunger to press the liquid material through the nozzle. The nozzle of the syringe can be modified by mounting a needle, available in different diameters for different volume and precision dispensing needs, see Table 4.5. The configuration of the "BLS syringe" depends on the used system layout, e.g. in the "Compact Lab Station" [6] or in the large "Integrated Lab Station" [7].

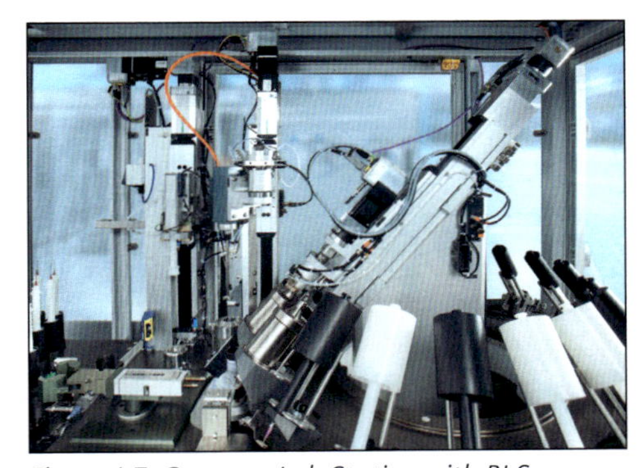

Figure 4.7: Compact Lab Station with BLS cartridges and dispensing carousel
Source: Füll Lab Automation

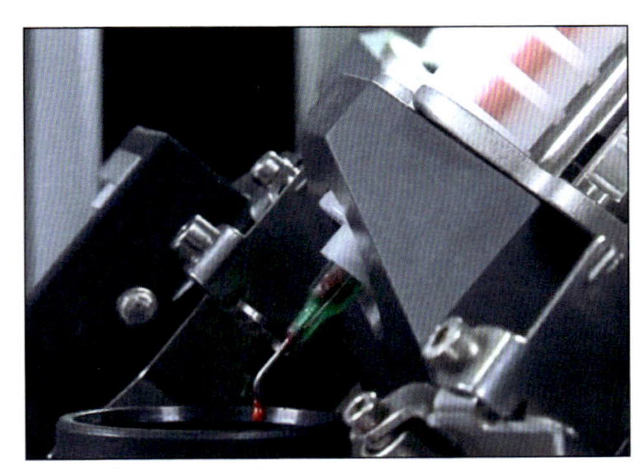

Figure 4.8: BLS syringe mounted on carousel during dispensing into formulation container
Source: Füll Lab Automation

Table 4.6: Line dispenser of Füll Lab Automation

Key characteristics	**Description**
Dispensing concept	Direct dispensing from feed vessel into formulation container
Material	From low viscous/medium viscous liquids viscosity range 0.2 to 500 mPas.With pt dispenser from low viscous to high viscous liquids supplied from feed vessel, viscosity range 0.2 to 30000 mPas, at room temperature, higher viscosities with heated container, tube and valve
Size of feed container	100 ml to 100 l (unlimited) With pt dispenser from 5 ml to 45 l
Storage capacity	Up to 12 line connections on one module corresponding to 12 feed containers. With pt dispenser up to 10 line connections corresponding to 10 feed containers on one module
Recommended dispensing range	10 mg to 500 g
Balance resolution	0.1 mg
Balance	On-deck, automatic dynamic weighing, volumetric dispensing possible

Figure 4.9: "Compact Lab Station" Source: Füll Lab Automation

Figure 4.10: Line dispenser in the "Compact Lab Station" supplied from feed vessels Source: Füll Lab Aut.

Figure 4.11: Mobile on-deck balance beneath liquid bulk dispensers Source: Labman Automation

Figure 4.12: Battery of 18 liquid bulk dispensers Source: Labman Automation

Table 4.7: Large liquid bulk dispenser in "Labman Paint Formulation System"

Key characteristics	**Description**
Dispensing concept	Direct dispensing into vessel on mobile on-deck balance
Material	Viscous liquids such as binders, viscosity range n/a
Size of feed container	2 l
Storage capacity	Max. 18 bulk dispensers with stirrers (6 pressurized and 12 peristaltic pumps equipped bulk dispensers)
Recommended dispensing range	0 to 190 g
Balance resolution	1 mg
Balance	Mobile on-deck, automatic dynamic weighing

A run can include one or several liquid dispenses from the set of syringes fixed on the carousel, Figure 4.7. The liquids are weighed into a sample cartridge, in which they can then be mixed using a stirrer. Despite to the compact design, there are 18 positions available on the carousel for dosing. Füll Lab offers the possibility of unlimited amount of different raw materials by a quick exchange of raw material/BLS syringes. A standard rack has 32 - 100 positions for BLS syringes. Several raw material racks are possible depending on the customized system design. In addition, the carousel can be equipped with double pt valves for liquids in pressurized containers (2 raw materials at each position) or powder containers.

The second option for liquid dosing is offered by Füll Lab with the "Line Dispenser", Figure 4.9 and 4.10. To do this, the formulation cartridge is placed in the holder on the scale below the "Line Dispenser". The "Line Dispenser" is then lowered over the formulation cartridge. In principle, with the "Line Dispenser" larger quantities could be dispensed compared with the "BLS syringe", see Table 4.6.

A modification of the "Line Dispenser" is also available. This enables the liquids to be dispensed while stirring, for example.

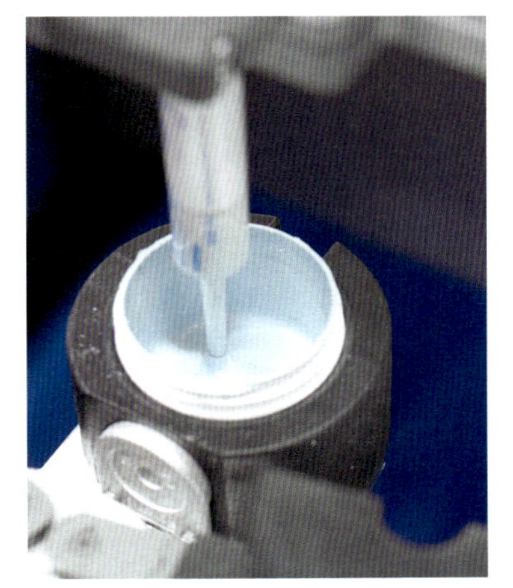

Figure 4.13: Dispensing of liquid from bulk dispenser into sample vessel positioned on on-deck balance
Source: Labman Automation

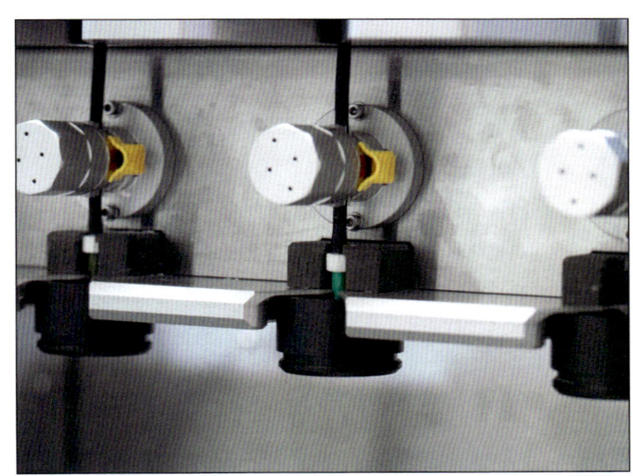

Figure 4.14: Pinch valves are responsible for the dispensing of the bulk liquids
Source: Labman Automation

Table 4.8: Mobile liquid dispenser in "Labman Advanced Formulation System"

Key characteristics	Description
Dispensing concept	Direct pressure dispensing into vessel
Material	Viscous liquids, viscosity range n/a
Size of feed container	2 l
Storage capacity	2 x 13, thereof 13 heatable
Recommended dispensing range	0 to 500 g
Balance resolution	10 mg
Balance	On-deck, automatic subtractive dynamic weighing

Figure 4.15: Transfer of the liquid dispensing unit from storage to formulation station Source: Labman Automation

Figure 4.16: Dispensing unit is placed on the on-deck balance Source: Labman Automation

Figure 4.17: Liquid dispensing unit with dispensing head at the front Source: Labman Automation

The Labman "Paint Formulation System" [8] offers various fully embedded dispenser types, within individual dispense stations. The system can be built with any number of dispense stations to accommodate the user needs. The dispensing of bulk liquids and slurries is done via peristaltic pumps or pressurized vessels with pinch valves into sample vessels. The "Paint Formulation System" of Labman can accommodate up to 18 x bulk liquid dispensers, Figure 4.11 and 4.12. The materials are stirred continuously throughout the run to ensure homogeneity.

The concept is based on one sample vessel moving by the 7-multiaxis robot from the storage onto the weighing position of the on-deck balance. This balance is positioned on a mobile table beneath the formulation platform. For each weighing step the balance is moved in x-direction to one of the 18 bulk dispensers, before moved in y-direction beneath the dispensing head,

After all large liquid volumes have been dispensed into the sample vessel, this one is transported to the next workflow step, e.g. small volume dispensing steps with the later described syringe dispensing, Figure 4.13 and 4.14 and Table 4.7.

The "Advance Formulation System"[9] is the next step up from the "Paint

Table 4.9: Small liquid syringe dispenser in "Labman Formulation System"

Key characteristics	**Description**
Dispensing concept	Direct pressure dispensing into vessel
Material	Viscous liquids such as additives and tinters, viscosity range n/a
Size of feed container	Approx. 50 ml
Storage capacity	2 x 16 syringes
Recommended dispensing range	10 mg to 5 g
Balance resolution	1 mg
Balance	On-deck, automatic dynamic weighing

Formulation System", see Table 4.8. This system uses new and leading technologies to deliver a robust platform that can be used in very flexible ways by its operators. Its flexibility allows it to run in ways that best optimise the samples being formulated. The 26 available dispensing containers with integrated dispensing heads are retrieved from the storage by the multi-axis robot and transferred to one of four formulation stations, Figure 4.15 and 4.16. There, the dispensing container unit is placed on the on-deck weighing device and tared. The liquid dispensing container gets slightly pressurized before dispensing, Figure 4.17. The dispensed quantity is measured subtractive and dynamically during the dispensing process.

The dispensing step directly into the formulation vessel could theoretically also done while stirring. For higher viscous material, the system has 13 heatable storage positions. This reduces the viscosity and makes the dispensing easier and more precise. The relatively heavy dispensing unit is connected horizontally to the mounting tool head of the multi-axis robot. During transfer processes, clear vibrations are recognizable, which indicate the load limit.

More critical, minor additions of tinters and additives are added via pressurized syringes with an accuracy of 3 milligrams. The prefilled syringes are in two racks on the automation platform[8]. Before the dispensing step, the syringe is moved to the syringe stirring station, where a non-contact magnetic stirring is performed.

After the sample is stirred up, the syringe is transferred into the syringe dispensing station, Figure 4.18, 4.19 and Table 4.9. The dispensing is performed directly into the sample vessel on a separate on-deck balance beneath the platform.

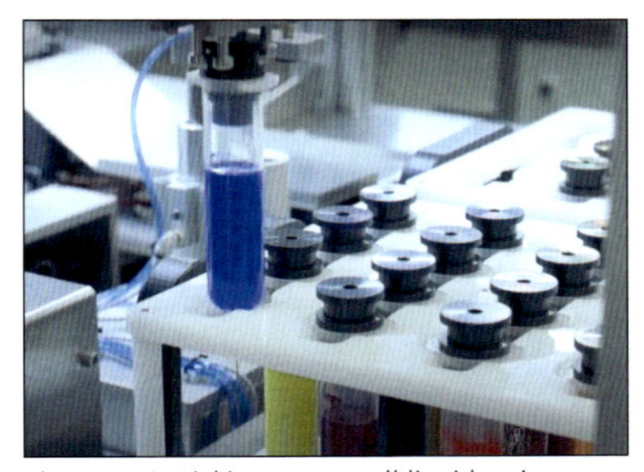

Figure 4.18: Picking up a small liquid syringe dispenser from the storage rack

Source: Labman Automation

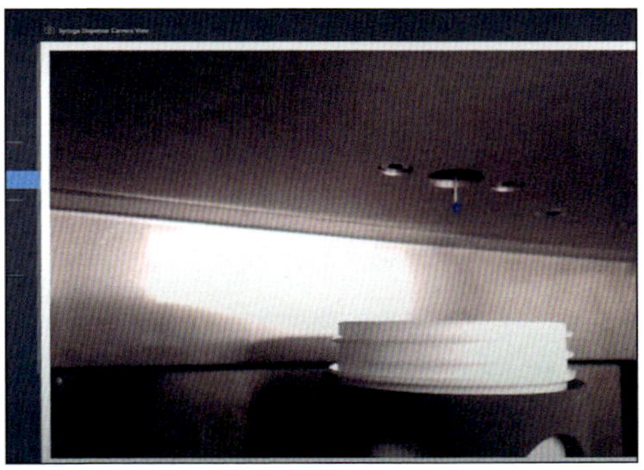

Figure 4.19: Fine dispensing of droplets through sringe dispensing needle into sample vessel

Source: Labman Automation

Table 4.10: "Flex Powder S" with GDU-Pfd of Chemspeed Technologies

Key characteristics	Description
Dispensing concept	Subtractive gravimetric weighing direct into vessel
Material	Powders, extrudates, etc.
Size of feed container	20 ml
Storage capacity	8 container multiples
Recommended dispensing range	1 mg to 2 g
Balance resolution	0.1 mg
Balance	Overhead analytical balance, automatic dynamic weighing

4.3 Gravimetric powder dispensing systems

Precise gravimetric dosing of powders is one of the most difficult tasks in laboratory automation. The difficulties encountered here are comparable to those with large silos on a production scale. Furthermore, not all powders are the same. There are very large differences in terms of particle size, particle shape, polarity in the form of hydrophilic or hydrophobic behavior, moisture, bulk density, etc. This has a strong impact on the formation of agglomerates and lumps, bridging in

Figure 4.20: Gravimetric dispensing unit for fine powder dosing called GDU-Pfd
Source: Chemspeed Technologies

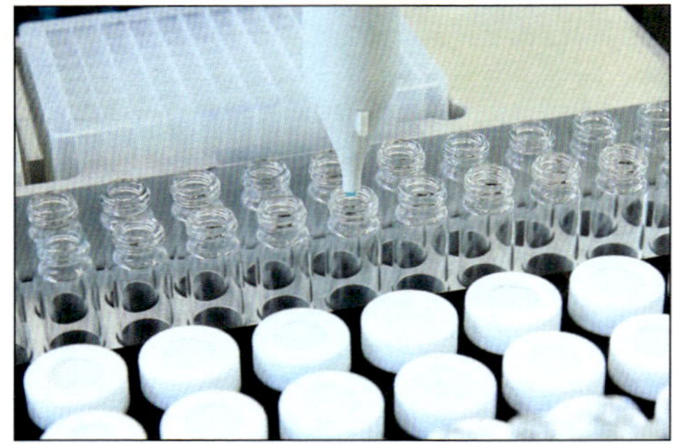

Figure 4.21: Fine dispensing of sticky powder into glass vials
Source: Chemspeed Technologies

Figure 4.22: Flex Powder M with GDU-P in a Flex-shuttle setup
Source: Chemspeed Technologies

Figure 4.23: Powder dispensing with overhead balance GDU-P direct into sample vessel located on the shuttle
Source: Chemspeed Technologies

Table 4.11: "Flex Powder M" with GDU-P of Chemspeed Technologies

Key characteristics	**Description**
Dispensing concept	Subtractive gravimetric weighing direct into vessel
Material	Powders, extrudates, etc.
Size of feed container	100 ml
Storage capacity	8 containers with 8 multiples, total 64
Recommended dispensing range	20 mg to 25 g
Balance resolution	1 mg
Balance	Overhead analytical balance, automatic dynamic weighing

the storage container, flowability, dust formation and static charging, which generally do not make dosing any easier. Another difficulty is that with very small samples, the amount of solid to be dosed is also much smaller. If only very fine particle sizes are assumed, this does not play a role. However, if sands, grains or pellets are dosed, then the smallest quantity that can be dosed is of course the weight of a particle. Under certain circumstances, precision may suffer as a result, or the tolerance range may have to be increased. As already described in Chapter 4.2, the manufacturers of laboratory automation systems offer different concepts for gravimetric powder dosing. Some concepts are reminiscent of production; the difficulty is, of course, miniaturization. The automation solutions on the next pages will be again discussed in alphabetical order.

Chemspeed's smallest powder dispensing offer for the paint and coatings industry is called "Flex Powder S"[10]. The gravimetric dispensing unit (GDU) in the configuration powder fine dispensing (Pfd) is used to perform automated, gravimetrically controlled dispenses of a wide range of solids in the sub-milligram to multigram scale. Unique is the concept of the overhead moving balance to the destination vessel, mounted to the overhead robot, Figure 4.20, 4.21 and Table 4.10. This is increasing speed and efficiency as the balance can move on a formulation platform from one destination vessel to the next.

The dispensing module mounted to the GDU picks the appropriate powder container from the storage rack, dispense the solids and powders gravimetrically in the destination vessel, and then place the powder container back to its storage position and picks the next one. The dispensing unit includes an analytical overhead balance and allows dispensing and simultaneous weighing of powders into the sample vessel.

The next larger size of dispensing containers, called "Flex Powder M"[11] with GDU-P containers, have 5 times the capacity of the GDU-Pfd containers, Figure 4.22 and 4.23. The automatic gravimetric dispensing system is preferred for medium amounts of solids and powders. It provides as the smaller sister system automatic dynamic weighing of dispensed solids – continuous weighing while dispensing. The recommended dispensing range of 20 mg to 25 g can vary depending on the bulk density of the powder materials.

Serial connection of solid dispensing containers for large quantities is available by the offered GDU-P container storage hotel. The additional storage provides space for 64 GDU-P containers. The powder containers will be transferred by an overhead robot from the storage to the rack transfer station, where the GDU-P dispensing tool can pick them up and move to the dispensing station. Then the subtractive gravimetric weighing is performed direct into the sample vessel positioned on the shuttle.

To obtain a continuous powder flow during dispensing, the GDU-P (as well as GDU-Pfd) container has an integrated extruder to avoid powder blocking. During opening the dispensing head

Table 4.12: "Flex Powder M" with "YODA" of Chemspeed Technologies

Key characteristics	**Description**
Dispensing concept	Direct dispensing into sample vessel on shuttle
Material	Any solid morphology (powders, extrudates, etc.)
Size of feed container	150 and 250 ml
Storage capacity	Up to 68 "YODA" container with 5 multiples, in total 340
Recommended dispensing range	10 mg to 50 g
Balance resolution	1 mg
Balance	On-deck, automatic dynamic weighing

by rotating or oscillating with the GDU dispensing head, the extruder is moved simultaneously. Different designs of extruders are available and can be specifically used depending on the powder properties. For the elimination of static charges an ionization system is installed nearby the dispensing location, Figure 4.24 to 4.26 as well Table 4.11.

A different concept of powder dispensing is applied by Chemspeed's "Flex Powder YODA"[12] module. It allows for quick, highly accurate dispensing of versatile solids from fine to coarse powders and of different particle shapes from round over plate to needle, flake or crumble appearance. The unique YODA dispensing technology is combined with an integrated balance beneath the dispenser. For target amounts of 10 mg up to 50 g, the YODA tool, integrated within the FLEX platform in this example, allows for hassle-free solid dispensing by pinching a nozzle, Figure 4.27 to 4.29 as well Table 4.12.

A medium sized to very large storage system allows hundreds of "YODA" containers to be accessed. The trays with these containers are in a small warehouse and moved by the elevator system to the transfer station. From there a multi axis robot grips the container and moves it to the dispensing station. During the transfer the bottle is turned by 180° which does at the same time loosen up the powder for easy powder flow during dispensing. Therefore, the "YODA"

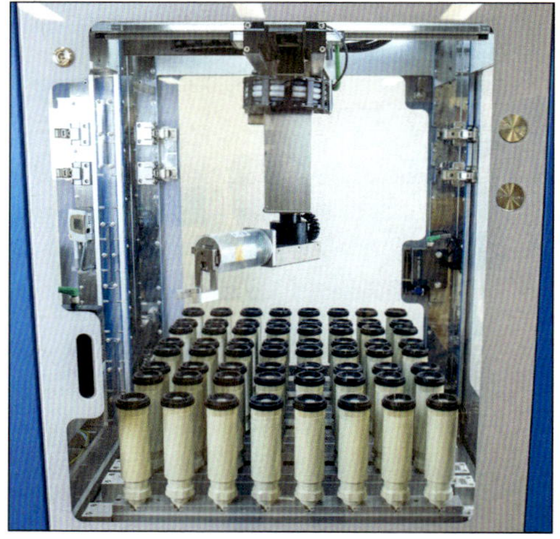

Figure 4.24: Flex Powder M with overhead robot for transfer of GDU-P containers

Source: Chemspeed Technologies

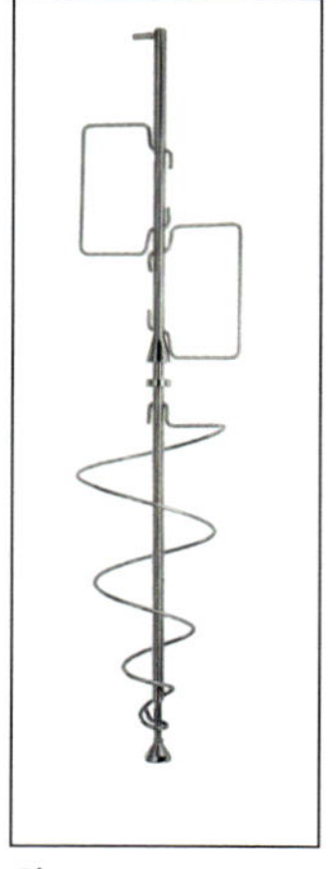

Figure 4.25: Extruder spindle for powder flow in container

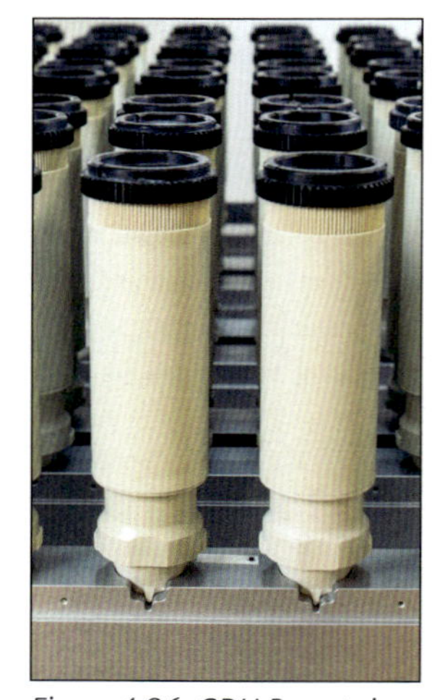

Figure 4.26: GDU-P container storage hotel

Source: Chemspeed Technologies

container does not need an extruder spindle as the GDU-P and -Pfd containers. This makes them easy to handle, e.g. filling of powder sample into this container, put the dispensing head with the dispenser ring on and close it for transport and dispensing. Thus, the "YODA" dispensing system is preferred used in quality control environment.

The biggest automatic gravimetric dispensing system for large amounts of solids and powders of Chemspeed is the "Flex Powder L", Figure 4.30 [13]. It shows its strength with frequently recurring and large dosing quantities of pigment and fillers. Formulation campaigns with high-solids paints, especially pigment pastes and interior wall paints, which often have a pigment volume concentration (pvc) of significantly more than 50 %, require large storage quantities in the laboratory automation system. Therefore, the "Powder L" container has a capacity of 5 l. On the carousel, which has place for 10 "Powder L" container, the needed container is moving towards the

Figure 4.27: "Flex Powder YODA" in a "Flexshuttle" setup Source: Chemspeed Technologies

Figure 4.29: "YODA" container in transfer by multi axis robot Source: Chemspeed Technologies

Figure 4.28: Powder dispensing with "YODA" container direct into sample vessel located on the shuttle Source: Chemspeed Technologies

Figure 4.30: "Flex Powder L" with 10 dispensing container on carousel

Source: Chemspeed Technologies

Table 4.13: "Flex Powder L" of Chemspeed Technologies

Key characteristics	Description
Dispensing concept	Direct dispensing into sample vessel on shuttle
Material	Any solid morphology (powders, extrudates, etc.)
Size of feed container	5 l
Storage capacity	Up to 10 dispensing containers
Recommended dispensing range	500 mg to 1 kg
Balance resolution	10 mg
Balance	On-deck, automatic dynamic weighing

dispensing station. If needed, these containers can be easily dismantled from the carousel and exchanged with new already filled container to continue or starting a new run.

At the dispensing station the powder is dispensed by automatic dynamic weighing directly on the sample vessel, which is located on the shuttle beneath. The shuttle with the sample vessel is parked on the on-deck balance, allowing a continuous weighing while dispensing, Figure 4.31. Despite the large size and weighing quantities, the on-deck balance has a resolution of 10 mg for high accuracy. The range of weighing is recommended from 500 mg to 1 kg. The transport of the powder within the container is ensured using an extruder spindle. For the powder flow at the dispensing had, Chemspeed has integrated a feeder screw, Figure 4.32 and Table 4.13.

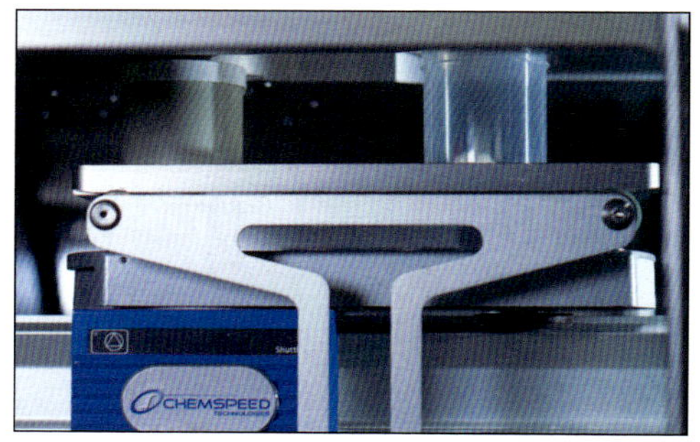

Figure 4.31: Direct dispensing into sample vessel located on the shuttle beneath the dispensing container Source: Chemspeed Technologies

Handling larger powder volumes need to have a safety concept versus contamination. During the powder dispensing an ionization system is eliminating the static charges. After the dispensing is finished, the carousel moves the dispensing container over the vacuum cleaning station to suck in possible adhering powder particles from the nozzle, Figure 4.33.

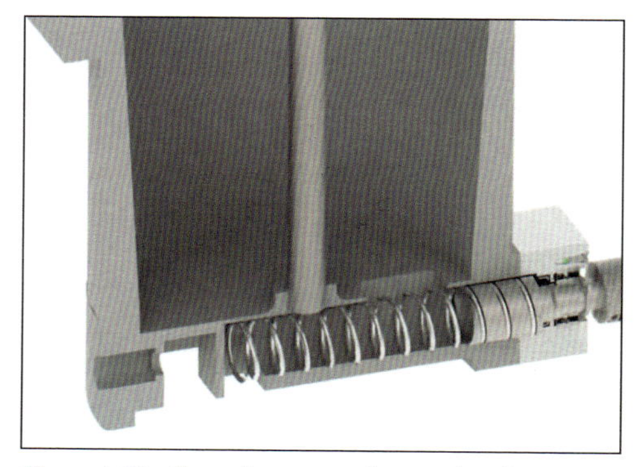
Figure 4.32: Extrusion screw for powder dispensing Source: Chemspeed Technologies

Figure 4.33: Ionization system and cleaning of dispensing head Source: Chemspeed Technologies

Table 4.14: Powder dispenser of Füll Lab Automation

Key characteristics	Description
Dispensing concept	Direct dispensing from powder container into sample vessel
Material	Powders, granules, fibres and milling beads
Size of feed container	250 and 500 ml
Storage capacity	Unlimited amount of different raw materials possible. Quick exchange of raw materials/powder containers possible. A standard rack has 32 positions for powder containers. Several raw material racks possible depending on customized system design
Recommended dispensing range	10 mg to 100 g
Balance resolution	0.1 mg
Balance	On-deck, automatic dynamic weighing

The portfolio of power dispensing systems at Füll Lab Automation is explained in the following section. There is a standard dispensing container available, which can be used in the "Compact Lab Station" (CLS) [6] as well as in the "Integrated Lab Station" (ILS) [7]. The size of the container is available in 250 and 500 ml, Table 4.14. The powder container can be mounted on the carousel in the "CLS", Figure 4.34 and 4.35. There is also the option of a separate tray in the "CLS" with a capacity of 32 powder containers. The powder is dosed by a rotating extruder screw, presumably adjustable in speed to control the precision of the dosing.

Figure 4.34: Powder dispensing container on carousel Source: Füll Lab Automation

In the "ILS", the tray with up to 32 powder containers can be easily loaded into the system using a mobile tray trolley. From there, the powder container is transferred by the handling robot to the dosing station. At the end of dosing, the powder container is moved back onto the tray. This means that the system can be unloaded at the end of the run without any great effort.

The UK based automation supplier Labman Automation offers different systems in size and concept. The small powder dosing system is based on the integration of the "Quantos" powder dosing system from the Swiss company Mettler Toledo [14]. The first step in sample preparation for analytical methods such as HPLC is the precise and correct weighing of substances. The "Quantos" was developed and introduced by Mettler Toledo for capsule filling, HPLC

Figure 4.35: Powder container on single dispensing station over on-deck balance Source: Füll Lab Automation

Table 4.15: Small powder dispenser of Labman Automation

Key characteristics	**Description**
Dispensing concept	Direct dispensing into sample vessel
Material	Powders
Size of feed container	125 ml
Storage capacity	Up to 24 on carousel
Recommended dispensing range	10 mg to 10 g
Balance resolution	1 mg
Balance	On-deck, automatic dynamic weighing

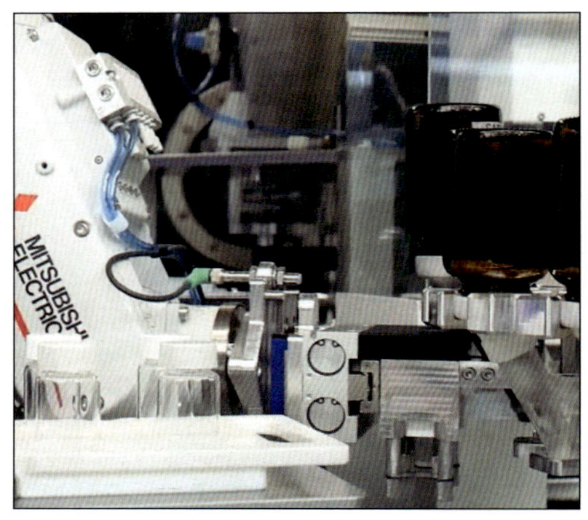

Figure 4.36: Robot picks up a small powder dispensing glass container from carousel storage
Source: Labman Automation

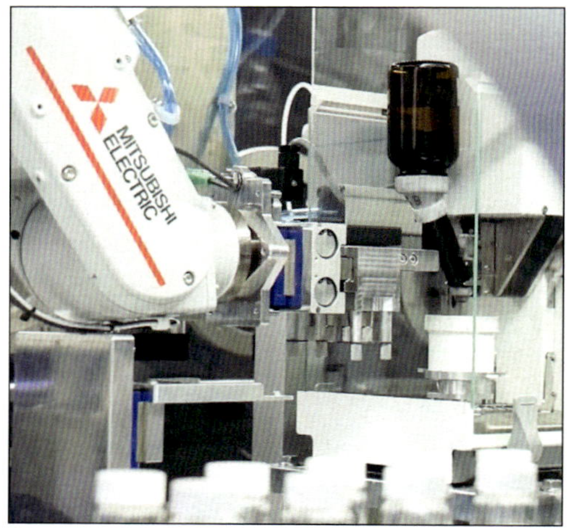

Figure 4.37: Small glass powder container at the dispensing station over on-deck balance
Source: Labman Automation

analysis, formulation or standard preparation. The automation development of the apperature from Labman resulted in the benchtop solution "Multidose" [15] with a collaborative robot and a storage carousel for 48 small dispensing bottles. The modified version for paint and coating applications allows 24 powder dispensing glass containers to be positioned on the storage carousel [16], Figure 4.36 and 4.37.

To start the dispensing process the multi axis robot picks up the desired powder dispensing glass container and transfers it to the dispensing station, which is slightly modified from the "Multidose". Labman has also a different configuration of this dispensing system, particularly a different storage system for the powder dispensing glass container, see Table 4.15. The powder is then dosed by vibrating and pulsing into the sample vessel beneath the powder dispensing glass container. Several powders can be dispensed in sequence. After the end of the dispensing, the sample vessel needs to be transferred to the next workflow step, allowing the next sample vessel to arrive for the next dispensing job.

Next solution of Labman is the powder dispenser in their "Paint Formulation System" [8]. This system offers just two powder hoppers. In the Figure 4.39 the dimensions look rather large with approximately 2 l capacity, but Labman states the powder feeder dispense is recommended up to 100 mg per sample, Figure 4.38 and 4.39 as well as Table 4.16.
The mobile on-deck balance moves with the sample vessel beneath the powder dispenser.

Table 4.16: Large powder dispenser in "Paint Formulation System" of Labman Automation

Key characteristics	Description
Dispensing concept	Direct dispensing into sample vessel
Material	Powders
Size of feed container	2 l (estimated)
Storage capacity	2
Recommended dispensing range	Up to 100 mg
Balance resolution	1 mg
Balance	Mobile on-deck balance, automatic dynamic weighing

To avoid spillage the sample vessel is lifted closer to the powder dispenser, before the gravimetric dispensing starts. Afterwards, the sample vessel moves together with the on-deck balance to the next workflow step.

The biggest capacity in number and flexibility for powder dispensing is offered by Labman with their "Advanced Formulation System", Figure 4.40 and 4.41 [9]. The user has the option of loading

Figure 4.38: Two large powder dispenser on "Paint Formulation System" Source: Labman Automation

Figure 4.39: Hopper dispenser over on-deck balance Source: Labman Automation

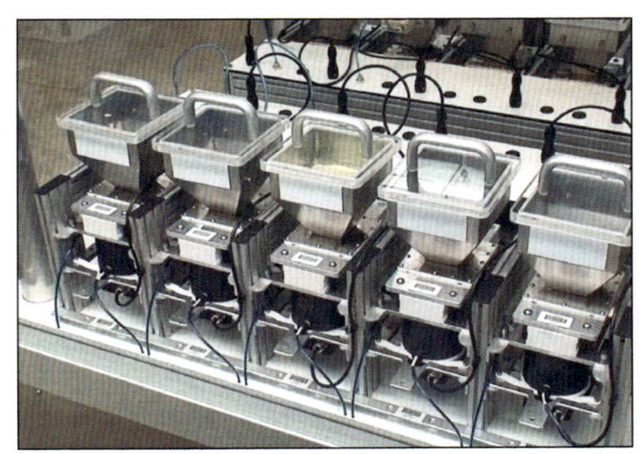

Figure 4.40: Powder hopper station in "Advanced Formulation System" Source: Labman Automation

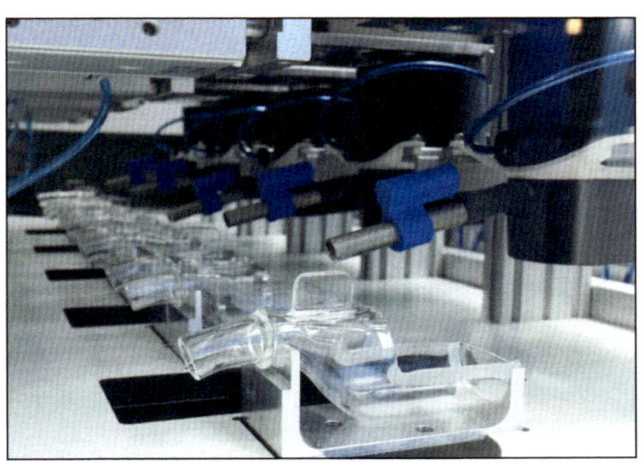

Figure 4.41: Auge feeder station below powder hopper station Source: Labman Automation

Table 4.17: Powder hopper with auge feeder of Labman Automation

Key characteristics	Description
Dispensing concept	Dispensing into auge feeder, then transferred to formulation vessel and poured in
Material	Powders
Size of feed container	1 l (estimated)
Storage capacity	10
Recommended dispensing range	0–20 g
Balance resolution	0.1 mg
Balance	On-deck, automatic dynamic weighing

10 powder hoppers with an estimated capacity of 1 litre into the system. The powder is dosed into an auge feeder, which is located underneath the powder hopper. A mobile on-deck scale lifts the auge feeder for the dosing process. A maximum of 20 g is precisely weighed in. The mobile on-deck balance then transports the auge feeder to the transfer station.

The multi-axis robot takes the auge feeder from the transfer station and transports it to the tilting mechanism for the formulation vessel. Once inserted there, it is tilted upwards and emptied into the formulation vessel by means of vibration shocks. Once it has been completely emptied, it is moved back to the dosing station under the powder hopper via the transfer station.

The auge feeder can then be reused without cleaning as only the same powder can be dispensed again during the same run, Figure 4.42 and 4.43. A lack of precision may occur in the case that not every particle is released from the auge feeder during emptying into the formulation vessel despite the use of vibration shocks.

4.4 Volumetric dispensing systems

Volumetric dispensing systems are crucial for precision and repeatability in various laboratory settings. They are reliable, versatile, and often easy to use, making them suitable for routine applications and high-throughput settings. Their ability to integrate with automated systems

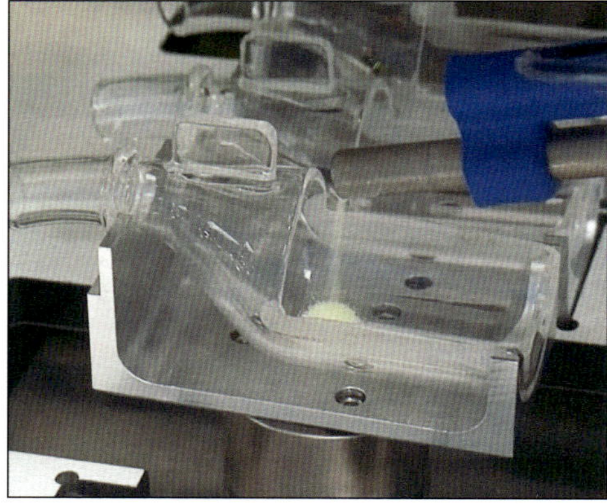

Figure 4.42: Dispensing of powder from hopper into auge feeder Source: Labman Automation

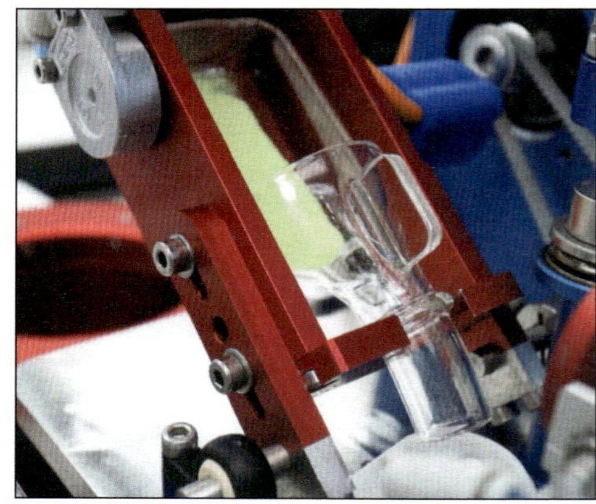

Figure 4.43: Flow of powder from auge feeder into paint formulation vessel

also makes them a foundational tool for modern laboratories, especially as labs shift towards higher automation and connectivity.

There are different types of volumetric dosing devices. They are widely used in fields requiring exact dosing, such as chemistry, pharmaceuticals, food and beverage testing, and quality control.

- Positive displacement pumps – commonly used in precision chemical synthesis, liquid chromatography, and pharmaceutical applications.
- Piston dosers – used in laboratories and production facilities for accurate dosing of solutions, especially in the food and beverage industries.
- Rotary volume pumps – often used for continuous dosing of liquids in industrial and laboratory applications.
- Pipetting systems – widely used in molecular biology, drug discovery, and genomics labs.
- Diaphragm pumps – useful in laboratories for chemical dosing, where control over dosing speed and accuracy is essential.
- Fixed volume pumps – suitable for routine tasks in laboratories like reagent dispensing, buffer preparation, and sample loading.

In summary, it can be said that volumetric dosing for laboratory automation systems for the formulation of paints and varnishes is more of a marginal phenomenon. However, many testing and analytical devices for the characterization and analysis use volumetric dispensing. The integration of such devices is further discussed in Chapter 7 "Technologies for testing and characterization". In automated production and filling lines for paints and coatings the picture is different. However, there are significant and therefore frequent applications, not part of this book, for:

- Pharmaceutical research and production
- Food and beverage testing
- Environmental testing
- Chemical synthesis
- Medical diagnostics

4.5 Literature

[1] Chemspeed Technologies, "Fully automated paints and coatings development and quality control", www.chemspeed.com/example-solutions/flexshuttle-paints-coatings/?s=Flexshuttle, 2024

[2] Chemspeed Technologies, "Flex Liquid S - Spartakus", www.chemspeed.com/media-center/video/flex-liquid-s-spartakus/?s=FLEX%20LIQUID%20S, 2024

[3] Chemspeed Technologies, "Flex Liquid M – Gizmo", www.chemspeed.com/media-center/video/flex-liquid-m/?s=FLEX%20Liquid%20M, 2024

[4] Chemspeed Technologies, "Flexshuttle QC with module Liquid L", www.chemspeed.com/media-center/video/flexshuttle-qc/?s=flexshuttle%20qc, 2024

[5] Füll Lab Automation, "BLS syringe", www.fuell-labautomation.com/products/bls-syringe, 2024

[6] Füll Lab Automation, "Compact Lab Station", www.fuell-labautomation.com/products/compact-lab-station, 2024

[7] Füll Lab Automation, "Integrated Lab Station", www.fuell-labautomation.com/products/integrated-lab-station, 2024

[8] Labman Automation, "Paint Formulation System", https://labmanautomation.com/portfolio/custom-system/paint-formulation-system, 2024

[9] Labman Automation, "Advanced Formulation System", https://labmanautomation.com/portfolio/custom-system/advanced-formulation-system, 2024

[10] Chemspeed Technologies, "Flex Powderdose", https://www.chemspeed.com/example-solutions/pick-and-dispense-of-powders-and-solids/?s=powder%20Pfd, 2024

[11] Chemspeed Technologies, "Flex Powder M", www.chemspeed.com/media-center/video/flex-powder-m/?s=Flex%20powder%20m, 2024

[12] Chemspeed Technologies, "YODA Dispenser", www.chemspeed.com/media-center/video/yoda-dispenser/?s=YODA, 2024
[13] Chemspeed Technologies, "Flex Powder L", www.chemspeed.com/media-center/video/flex-powder-l/?s=Powder%20L, 2024
[14] Mettler Toledo, "Solutions for Laboratory Weighing Automation", www.mt.com/ch/en/home/products/Laboratory_Weighing_Solutions/Weighing-Automation.html, 2024
[15] Labman Automation, "Multidose", https://labmanautomation.com/portfolio/products/multidose, 2024
[16] Labman Automation, "Formulation system with glovebox", https://labmanautomation.com/portfolio/custom-system/formulation-system-with-glovebox, 2024

5 Automated formulation

In paint labs, mixing devices are crucial for achieving consistent fineness of grind, colour, and other quality criteria in formulations. These devices ensure homogeneity in paint mixtures by effectively combining pigments, fillers, solvents, binders, and additives [1]. This allows the comparison of different raw materials and their effect onto paint and coating formulations. Regardless of whether production or laboratory scale, the most important universally applicable dispersion methods of manual workflows can also be found in laboratory automation. Preference is given to methods that do not require cleaning. As a result, consumables are increasingly being used. This also prevents contamination from sample to sample. Based on the before described facts, the use of dual asymmetric centrifuges (DAC) in laboratory automation is preferred by automation manufacturers. As a rule, this is also easy to reconcile, as high-throughput systems are generally used to screen raw materials and formulations. Fine tuning, i.e. the last 5 %, requires larger batch sizes so that more technical application tests can be carried out. Furthermore, the transfer or upscaling is also closer to practice when working with classic manufacturing methods. The following list provides an initial overview of dispersion methods that can be used depending on the paint and coating system and its viscosity.

- High speed disperser
- Ball mills
- Planetary mixers
- Dual shaft mixers
- Propeller mixers ("stirrers, agitator")
- Gyroscopic mixers
- Three roll mills
- Ultrasonic mixers
- Magnetic stirrers
- Static mixers
- High pressure homogenizers
- Vacuum mixers

Table 5.1 provides an overview of the preferred dispersion methods, and the specific use of automation manufactures.

As explained in the previous chapters of this book, the choice of automation providers for the paint and coatings industry is currently still very limited. In other chemical industries, such as life science and material science, there are more specialized providers, especially for sample preparation. For this reason, the focus in the following sub-chapters is once again on the three companies Chemspeed, Füll Lab and Labman.

5.1 Stirrer systems

The German automation manufacturer Füll Lab uses agitators with a stirring geometry that is a mixture of a propeller and a toothed disk for mixing and pre-dispersing. This system is used in the "CLS" of Füll Lab, see Figure 5.1. Liquid and solid raw materials are added, partly without and partly with stirring. The dimensions of the agitator and the diameter of the formulation contain-

Table 5.1: Overview matrix of mixing and dispersing systems per supplier

Mixing/dispersing system	Chemspeed Technologies	Füll Lab Automation	Labman Automation
Stirrer	X	X	X
High speed disperser	X	X	X
Dual asymmetric centrifuge (DAC)	X	X	X
Formulation reactor	X		
Oscillating shaker	X		

er in Figure 5.2 are not ideal. Practical experience shows that the selected diameter of the agitator is either too large or the diameter of the formulation container is too small[2]. A cleaning station for the stirrer is in the immediate vicinity of the stirring station. The same agitator concept with agitator geometry and size dimensions of the "CLS" is also used in the "ILS". There, however, it is only used for pre-dispersion before the actual dispersion takes place in a dual asymmetric centrifuge.

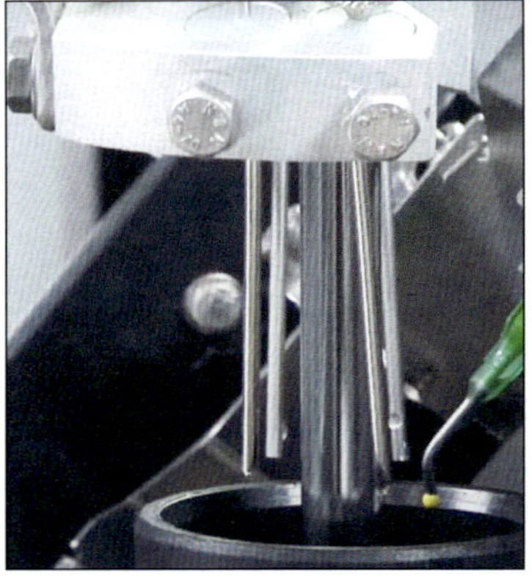

Figure 5.1: Dispensing of liquids during stirring Source: Füll Lab Automation

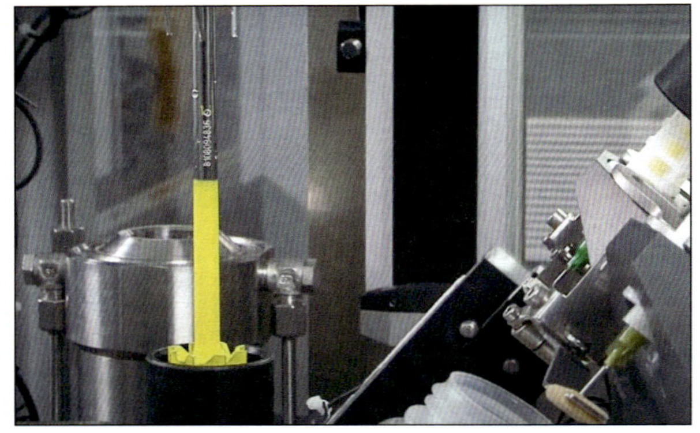

Figure 5.2: Agitator with stirrer geometry and formulation cartridge beneath, cleaning station in the back Source: Füll Lab Automation

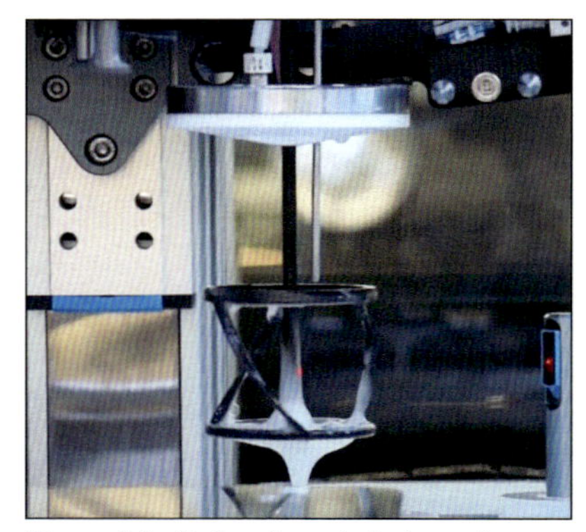

Figure 5.3: Helix stirrer in Labman's Formulation Engine Source: Labman Automation

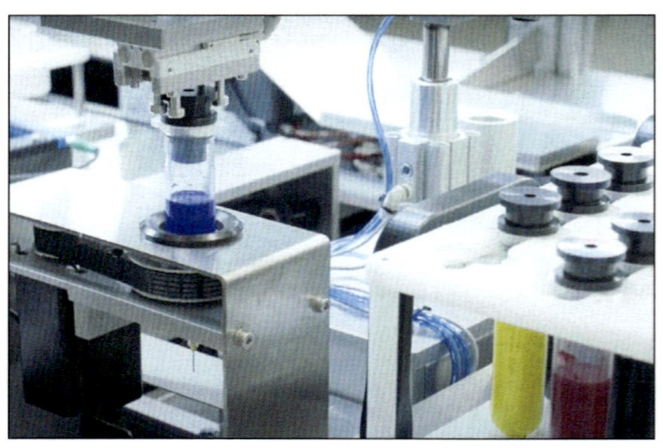

Figure 5.4: Magnetic stirrer for homogenizing of small liquid syringes Source: Labman Automation

Labman's portfolio also includes the "Formulation Engine"[3], touted as the next-generation modular formulation system[4]. The "Formulation Engine" is equipped with three helix mixers inclusive pH/temperature probe and optional a homogenizer (for high shear mixing) among other features.

There are no other uses of stirrers for premixing or dispersing paints and coatings. However, there are indirect applications of stirrers during the production of paints and coatings in lab automation. There are many raw material containers that are stirred during storage or shortly before dispensing. Different stirring concepts come into question here. Classic agitators are integrated in many liquid raw material containers, e.g. the mobile dispensing units (MDU) for liquids of Labman, see Figure 5.5. The different sized liquid containers in the "Flex Liquid M" module from Chemspeed are also equipped with stirrers, depending on the configuration, see Figure 5.6.

5.2 High speed disperser systems

Significantly higher shear forces than with conventional stirrers are achieved with high-speed stirrers known as dissolvers. These are also necessary to finely disperse pigments and fillers in a liquid medium. For this purpose, agglomerates of the solids must be converted into their primary particles to achieve their full functionality for optical, mechanical and chemical properties. The high-speed stirrers can be equipped with different stirring geometries. Traditionally, toothed disks are used without additional shear forces. For higher shear forces grinding aids such as glass or zirconium beads can be added, especially for low-viscosity systems. However, the dissolver can also be combined with scraper(s) or anchor agitators. These are known as dual-shaft mixers.

Figure 5.5: Stirred mobile dispensing units for liquids
Source: Labman Automation

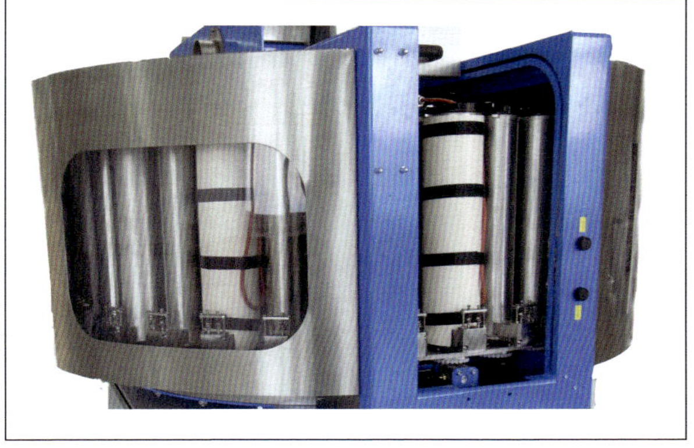

Figure 5.6: Stirred liquid dispensing containers in Flex Liquid M
Source: Chemspeed Technologies

Füll Lab offers for dispersing paints and coatings overhead stirrer(s) in combination with a stirrer cleaning station operated with solvents and/or water. The user can choose from a portfolio of standard stirrer or high speed/high shear mixer with stirring velocities between 0 and 20,000 rpm. In addition, different stirrer disk geometries are available such as saw tooth stirrer, dissolver disk, or rotor stator mixer. Stirrer geometries can also be adjusted to fulfil the requirements of a specific workflow. The dimensions of stirrer disks versus formulation vessel/cartridge are selected depending on the size of the formulation vessel and the workflow requirements, Figure 5.7.

Labman, for example, has equipped its "Advanced Formula-

Figure 5.7: Battery of high-speed dispersers
Source: Füll Lab Automation

Figure 5.8: Mobile dispersion unit incl. dual-shaft mixer and on-deck balance on trolley *Source: Labman Automation*

Figure 5.9: Battery of 4 liquid dispersion units
Source: Labman Automation

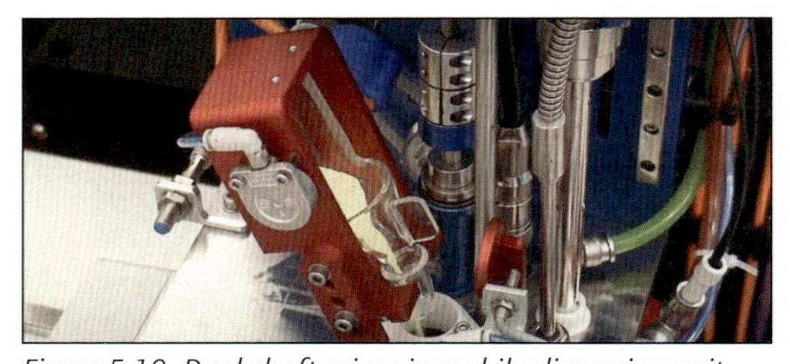

Figure 5.10: Dual-shaft mixer in mobile dispersion unit
Source: Labman Automation

tion System" with dual shaft mixers. The example shown has a total of four mobile dispersing units on trolleys. This indicates that the system can be configured with more or fewer dispersion units. The dispersion unit is equipped with a dual shaft mixer consisting of an anchor stirrer and a propeller stirrer. The sample container can be moved sideways by means of a slider. This enables the dispersion unit to be loaded and unloaded with the formulation vessel. As soon as the formulation vessel is unloaded from the dispersion unit, the dual-shaft mixer can be used to reach the wash station below, see Figure 5.8 to 5.10. Detailed information about dispersion conditions such as speed range, torque, stirrer geometries are not available.

Chemspeed's modular concept offers three fully integrated dissolvers in the "Flex Dispersion", see Figure 5.11. However, the module can be upgraded to a total of 5 dissolvers. The equipment includes a toothed disk with a diameter adapted to the formulation container as standard, see Figure 5.12. Alternatively, a "Teflon" disk can also be selected, as is common in vertical bead mill setups, for example, see Figure 5.13. Chemspeed's "Flexshuttle" concept also offers the gravimetric dispensing of glass beads and/or zirconia beads during the dispensing steps before the dispersion process. The freely controllable speed range of the dissolver extends up to 20,000 revolutions per minute. The speed ramps and duration of dispersion can be adapted to the workflow. The maximum torque of the dissolver is 0.8 Nm. Mounting on the vertical axis allows height adjustment and adaptation to the container and filling height of the formulation. Each individual dissolver has a PT100 temperature sensor for measuring the temperature during dispersion. Depending on the configuration, a fixed number of liquids can also be added during dispersion. For temperature control of the formulation vessel, the vessel holder is equipped with a heat jacket supplied by a cryostat. A slider is used for loading and unloading the formulation vessel in the same way as Labman and a linear axis system is also available for transferring the formulation vessel to and from the shuttle. The slider solution also provides access to the active washing station below for the dissolver. The "Flex Dispersion" system can be used not only for aqueous systems, but also for solvent-based systems. That is why the washing station also contains an F90 safety cabinet.

To operate the "Flex Dispersion" as a vertical bead mill with "Teflon" disk, it requires the controlled addition of beads. Precise gravimetric dispensing is achieved with the "Flex Bead Dispensing" module, see Figure 5.14. Depending on the configuration, either glass beads or zirconium beads with a diameter of 0.4 to 2.0 mm can be dispensed. Each storage container holds 25 litres of beads. Depending on the formulation vessel size, the dosage is between 10 and 100 g. The dispensing accuracy is 200 mg, see Figure 5.15. To prevent beads splashing a special enhancement floor containment has been integrated.

After dispersion, the beads must of course be removed from the coating. This is done

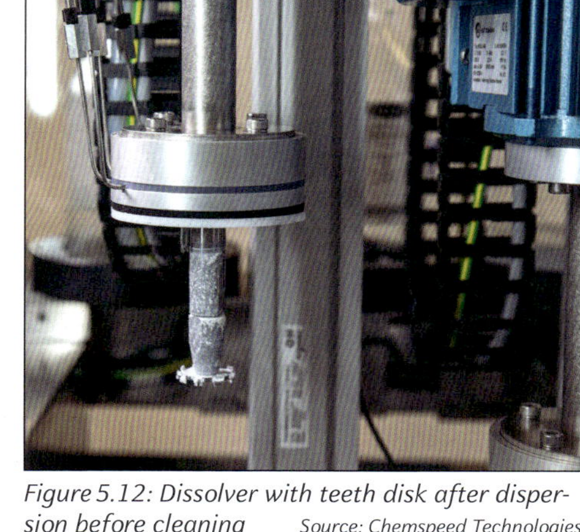

Figure 5.12: Dissolver with teeth disk after dispersion before cleaning Source: Chemspeed Technologies

Figure 5.11: "Flex Dispersion" with heat-jacked formulation vessel holder Source: Chemspeed Technologies

Figure 5.13: "Flex Dispersion" with 4 Teflon blade equipped dissolvers Source: Chemspeed Technologies

using a specially developed screen separation process in the "Flex Filtration" module. This is faster and more efficient than the manual procedure with an average of two to four times higher yield. For this purpose, a filter adapter with a built-in sieve and the target glass container and lid are pressed firmly into the formulation container using a multi-axis robot. The integrated lift storage holds 320 filter adapters for the filtration process, see Figure 5.16. The robot, which can grip the larger formulation beaker as well as the smaller target vessel with its multi gripper, ensures the transfer into a centrifuge, see Figure 5.17. The coating is pressed through the sieve at more than 1,000 rpm and the beads are retained. The containers and filter adapter are then transferred from the centrifuge back to the workstation. Here, the formulation container with the filter adapter is unscrewed from the glass container, see Figure 5.18. The filter adapter (consumable material) is discarded together with the glass beads. Zirconium beads are retained and collected for reuse after cleaning. The coating material in the glass container is weighed back on the on-deck balance with 10 mg resolution so that the content is known for the next workflow step (let down). Shuttles are used to transport the glass containers with the dispersed coating material without beads to the next workflow step.

5.3 Dual asymmetric centrifuge systems

All suppliers of laboratory automation systems use dual asymmetric centrifuges for dispersion. These are ideal for screening of raw materials and developing formulations. The advantages are obvious:

- No cleaning of dispersing unit
- No contaminations
- Only one vessel per sample
- Deaeration

Some of the models used by the dual asymmetric centrifuge manufacturers have different features. However, they all have the option of working with adapter inserts to use differently dimensioned formulation vessels. For automation laboratory systems, the smallest DAC available is typically a model with a batch load of 400 g. Smaller systems cannot be used, as they have not the space to integrate a forced stop. Without a forced stop of the vessel holder, the load-

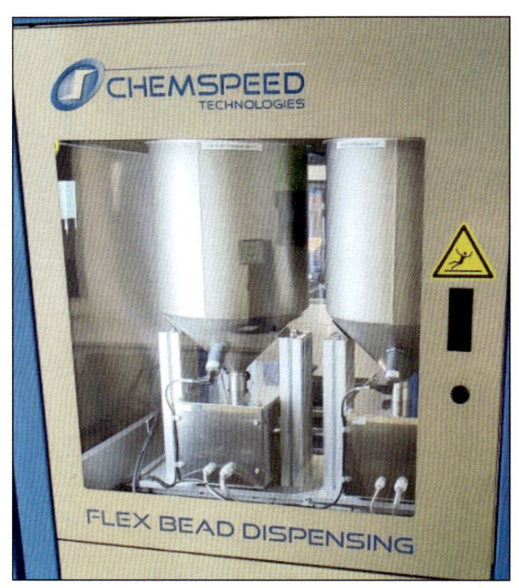

Figure 5.14: "Flex Bead Dispensing" module with double feed container Source: Chemspeed Technologies

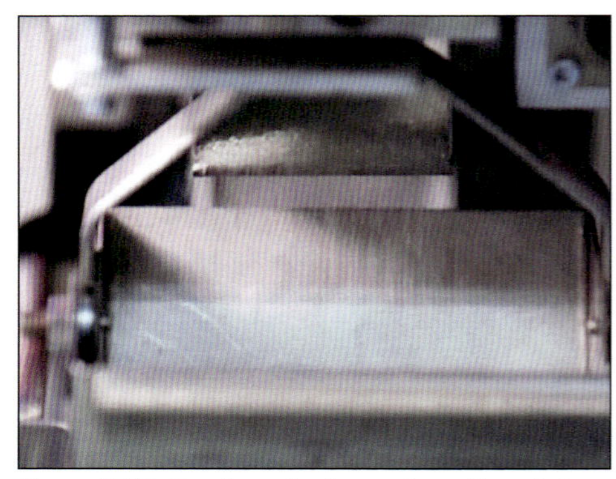

Figure 5.15: Gravimetric dispensing of beads into formulation vessel Source: Chemspeed Technologies

ing and deloading of the formulation container cannot be performed with a robot. Another major advantage of DAC's is their deaeration behaviour. The paints and varnishes dispersed with them no longer contain any air or foam. This means that the first tests and applications such as drawdowns can be carried out immediately after production without disturbing effects in the coating film.

Füll Lab also makes this their own by dispersing in a DAC their cartridge-shaped formulation container, part of their "Integrated Lab Station", see Figure 5.19. Mechanically, this works without any problems, but the geometric ratios of diameter to height are not particularly ideal, if the full available filling height is used. A lower height or wider diameter would improve the dispersion result. Therefore, depending on the viscosity, this works differently well with their setup. Füll Lab decided to integrate the "Speedmixer Smart DAC 400" of Hauschild Engineering, which is their smallest available DAC with a forced stop.

Of course, Labman has also been using Hauschild Engineering's DAC technology for years. As can be seen in the figures, the "Speedmixer DAC 400" with one vessel position is installed alongside the "Speedmixer" version Smart DAC 400, see Figures 5.20 and 5.22. However, the DAC 400 speed mixer can also disperse more than one sample at a time. With the appropriate insert adapter, three formulation vessels can be loaded simultaneously in the example shown in Figure 5.21. The

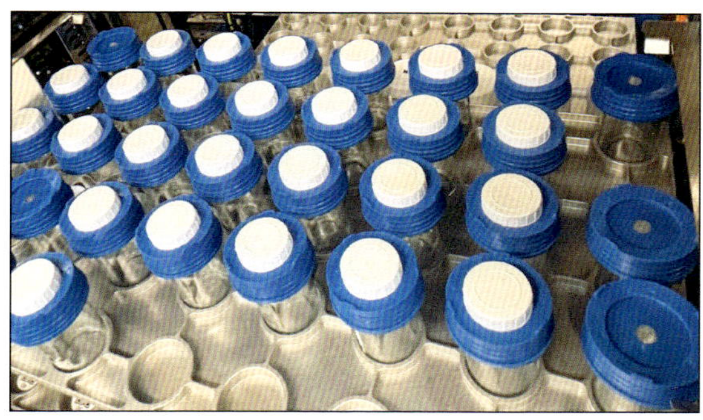

Figure 5.16: Filter adapter screwed on glass jar and lid
Source: Chemspeed Technologies

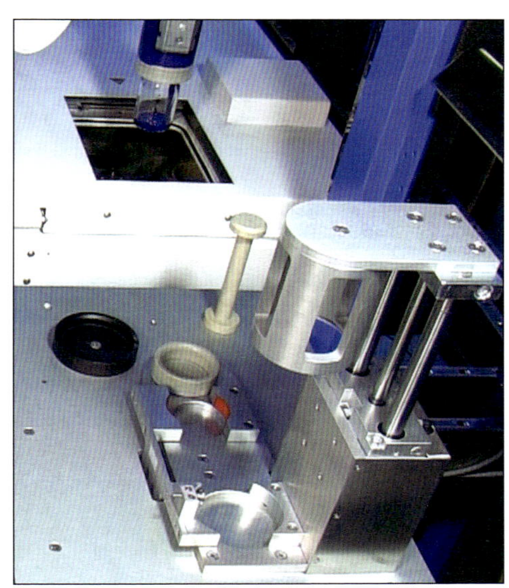

Figure 5.17: Inserting the container with filter adapter into the centrifuge using a multi-axis robot
Source: Chemspeed Technologies

Figure 5.18: Formulation vessel with retained beads after filtration and glass jar with only coating material
Source: Chemspeed Technologies

adapter with the three samples is then inserted into the “Speedmixer” using the multi-axis robot. It is only necessary to ensure that the gross weight loading is not exceeded. The new Smart DAC version also allows two vessel positions as an option. In addition, the new Smart DAC offers variable control of the counter-rotation, i.e. the counter-rotation can be increased or decreased independently of the main rotation. This development has already been successfully launched on the market by the Japanese DAC manufacturer Kurabo. This is particularly advantageous when mixing solids and liquids. Different and low counter-rotation of the formulation vessel compared to the main rotation prevents pigments and fillers from being carried into the lower edge area of the formulation vessel without being wetted and remaining there as "dead" lumps until the end of the dispersion process. This was precisely the weakness of the first DAC mixers and their use in laboratory automation systems.

The Swiss lab automation manufacturer Chemspeed offers their customers different DAC models according to their functional requirements. DAC’s from Hauschild Engineering, see Figure 5.23,

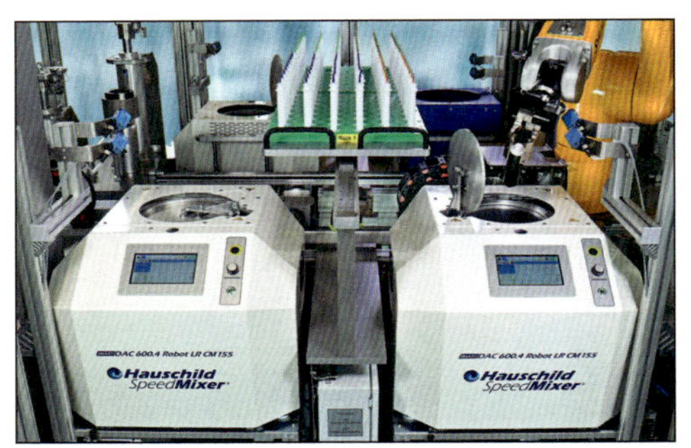

Figure 5.19: Two Speedmixer during loading and operation Source: Füll Lab Automation

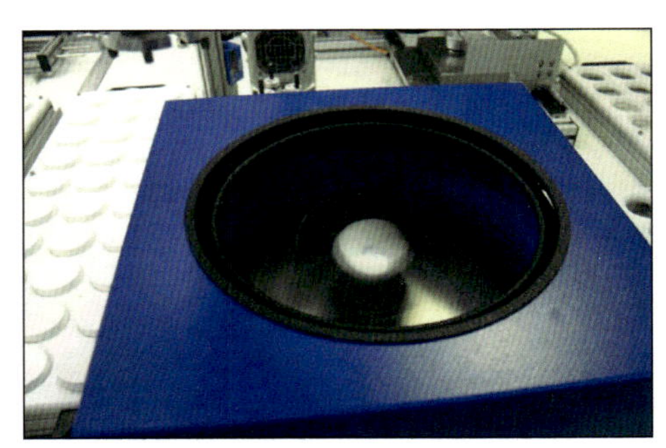

Figure 5.20: Mixing of formulation vessel in “Speedmixer DAC 400” Source: Labman Automation

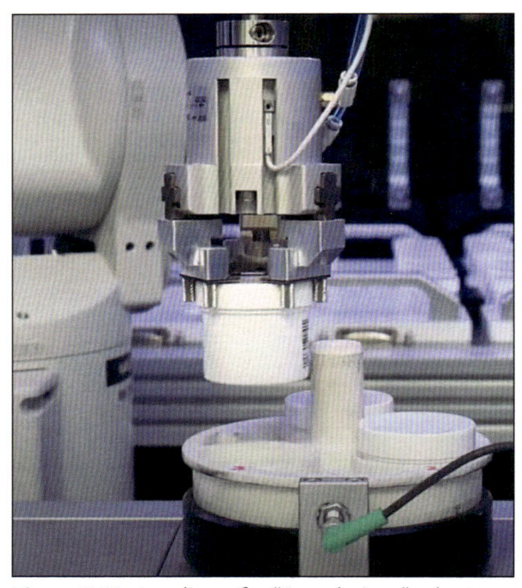

Figure 5.21: Loading of a “Speedmixer” adapter with a capacity of 3 formulation vessels Source: Labman Automation

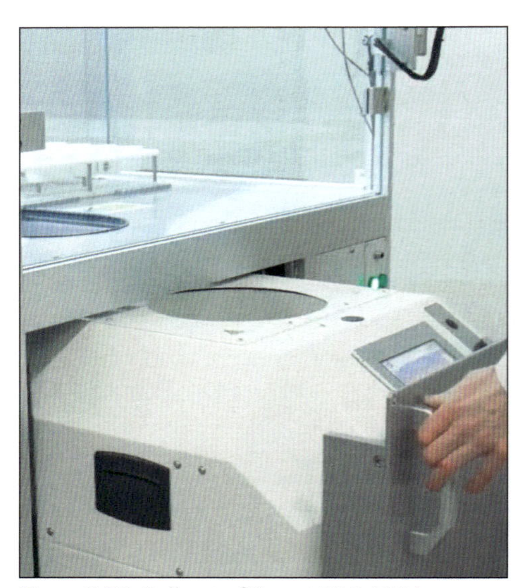

Figure 5.22: Pull out of Smart DAC 400 in Labman’s “Paint Formulation System” for better maintenance access Source: Labman Automation

and Kurabo can be integrated into the systems. A separate counter rotation and main rotation during the dispersion offers many advantages. In the meantime, Chemspeed has integrated the "Mazerustar KK-400W" frequently, see Figure 5.24 and 5.25. The formulation vessel positions offset from the axis of rotation also require lower speeds to achieve the same dispersing effect as operation with a DAC with only one formulation vessel position centrally on the axis of rotation. Another advantage is the simultaneous operation of two formulation vessels. If only one sample is to be dispersed, a corresponding counterweight must be taken from the adjacent storage and loaded into the DAC. If the throughput is still insufficient, several DAC's can also be configured in one module.

Figure 5.23: Double "Speedmixer DAC 400" setup in "Flexshuttle" Source: Chemspeed Technologies

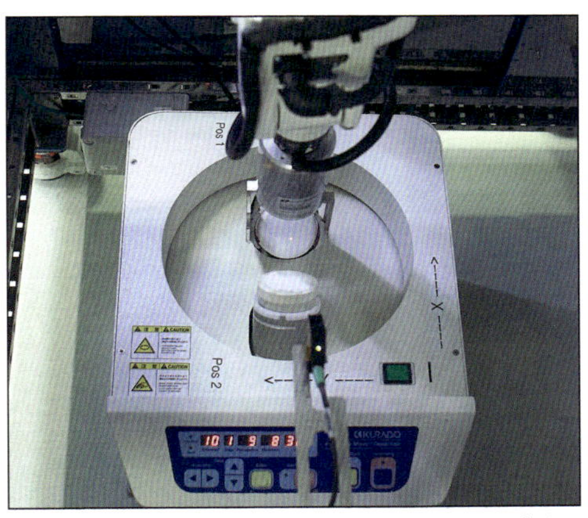

Figure 5.24: Insertion of 2 formulation vessels in Mazerustar KK-400W Source: Chemspeed Technologies

5.4 Formulation reactor

The "Formax Paints and Coatings" [5] is a modular robotic platform enabling automated high-quality formulations for paints and coatings [6-8], see Figure 5.26. It has been designed with flexibility allowing the modification and/or adding new tools, racks, vessels/blenders, application and testing tools at any time with the choice of up to 70 tool-features, see Figure 5.29. The workstation has besides different dispensing tools for liquids and powders, as discussed earlier in Chapter 4, a configurable setup of formulation reactors (formulation vessels) with either 100 ml or 1 l sizes. The system can be equipped with a maximum of 4 formulation blocks. Each formulation block can host either 6 formulation vessels of 100 ml or 3 formulation vessels of 1 l size, see Figure 5.27. The formulation blocks can be configurated on the system, allowing also a mix of 100 ml and 1 l formulation vessels, see Figure 5.28.

The special feature of the formulation reactor lies in the drive of the stirring units. As there is a lower and upper stirring unit, two separate motors are also required for individual control. For this reason, there are different dispersing

Figure 5.25: Flex Mix module with 4 Mazerustar KK-400W Source: Chemspeed Technologies

tools for the formulation vessel available, e.g. a teeth disk or rotor-stator. These are operated from the bottom motor with up to 6'000 rpm, even during gravimetric of highly viscous materials and powders. The maximum operational viscosity depends on the stirrer but is between 125 to 250 Pas. The top motor drives the scrapers, which are rotating with 20 to 200 rpm along the wall of the formulation vessel to ensure a proper mixing of all dispensed materials. Each formulation vessel can be individually controlled for speeds, temperature and viscosity by in-situ measurements. After the

Figure 5.26: "Formax" with 24 formulation vessels each of 100 ml size Source: Chemspeed Technologies

Figure 5.27: Formulation block with 6 formulation and feed vessels Source: Chemspeed Technologies

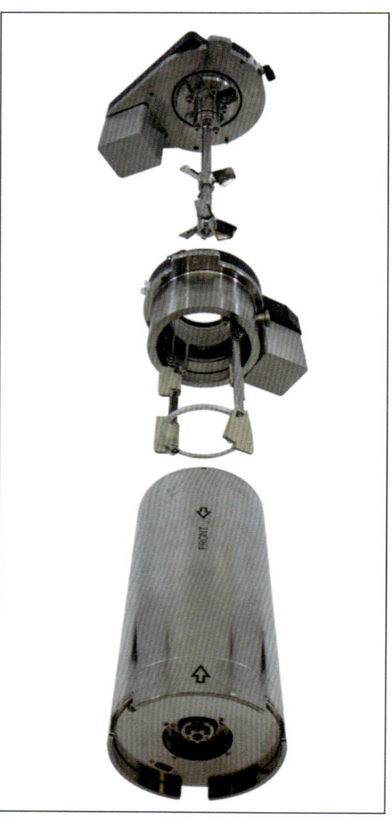

Figure 5.28: Formulation vessel of 1 l with configurable stirrers and blades Source: Chemspeed Technologies

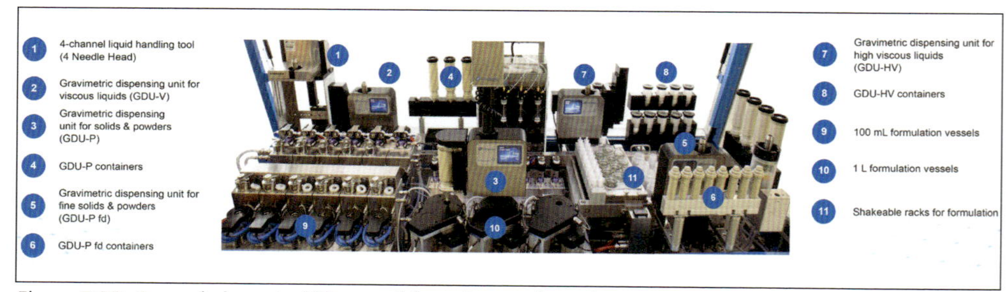

Figure 5.29: Example layout of "Formax" for paints and coatings Source: Chemspeed Technologies

formulation is completed, the formulation vessels need to be removed from the workstation and manually emptied as well as cleaned.

5.5 Oscillating shaker

Another option for dispersion is offered by Chemspeed in the "Flex Mix" module ("Lau Shaker"), see Figure 5.30. The principle was originally based on the invention of the US paint shaker "Red Devil" decades ago. It was later optimized with the development of the "Skandex Mixer". In the "Red Devil", the sample position and thus the influence of the amplitude of the shaking movement had a strong effect on the quality of the dispersion. The advantage of the "Skandex Mixer" lies in the elimination of the influence of sample positioning. The oscillating movement produces identical dispersion results regardless of the sample positioning. Today, "Skandex Mixer" principle is commonly used in tinting machines at DIY stores to mix colour concentrates into a white base colour paint. The German company LAU offers a slightly modified version of this concept for laboratory use. Several containers are loaded into a rack, which is then inserted and fixed in the "Disperser DAS H – System LAU" . The number of formulation vessels varies with the size of the container. The dispersing effect is controlled by the duration of operation. The maximum speed is noted with 1,450 rpm. The principle has been further modified for the automation purpose by Chemspeed. This mainly concerns the automatic loading and unloading as well as the fixing of the rack in the oscillating shaker. The configuration allows up to 25 glass vessels with a size of 100 ml to be operated per mixer, see Figure 5.31. The dispersion process is most frequently used for low-viscosity paints, which are therefore also mixed with beads. In the setup shown [9], a total of 4 "LAU" dispersers were integrated into the module "Flex Mix", which can therefore disperse up to 100 glass vessels simultaneously. Finally, it should be mentioned that, as with the DAC concept, the "LAU Disperser" does not require a cleaning station due to the use of closed glass vessels and therefore no contamination is possible.

Figure 5.30: Overview of "Flex Mix" with 4 "LAU" dispersers
Source: Chemspeed Technologies

Figure 5.31: Loading of a fully filled rack with 25 glass vessels into "LAU" disperser
Source: Chemspeed Technologies

5.6 Literature

[1] BROCK, T.; GROTEKLAES, M.; MISCHKE, P.; "European Coatings Handbook", p. 229ff, Vincentz Network, Hannover, 2010
[2] BROCK, T.; GROTEKLAES, M.; MISCHKE, P.; "European Coatings Handbook", p. 241ff, Vincentz Network, Hannover, 2010
[3] Labman Automation, "Formulation Engine", https://labmanautomation.com/portfolio/custom-system/formulation-engine, 2024
[4] Materials Innovation Factory at University of Liverpool, "Unparalleled technology", www.liverpool.ac.uk/materials-innovation-factory/facilities/, 2024
[5] Chemspeed Technologies, "FORMAX", www.chemspeed.com/example-solutions/formax-paints-and-coatings/?s=formax%20p, 2024
[6] VLCI, "High Throughput – Free your mind and increase knowledge", https://vlci.biz/high-throughput/, 2024
[7] KILPELÄINEN, V.; GUTIERREZ, A.; VAN LOON, S.; "Raising the barrier to rust", p. 26ff, European Coatings Journal, 04/2012
[8] VAN LOON, S.; FRICKER, B.; KOTHER, F.; "Optimizing Coating Performance via Predictive Compatibility Parameters of Carbon Blacks and Dispersants", https://coatings.specialchem.com/tech-library/article/carbon-blacks-dispersion-optimized-coating-performance, SpecialChem, April 23, 2021
[9] Altana press release, "ALTANA sets new standards in Wesel: digital BYK laboratory is unique worldwide", www.altana.com/press-news/details/altana-sets-new-standards-in-wesel-digital-byk-laboratory-is-unique-worldwide.html, 3rd May 2022

6 Application technologies

Paint application technologies have evolved significantly to improve efficiency, surface coverage, aesthetics, and environmental compliance. These technologies serving to various industries, including automotive, construction, aerospace, and consumer goods [1-4]. The specific coating technologies are generally applied in laboratory environment after the screening for advance application performance and judgement and, of course, in practice in the industries. Drawdown and spray processes are predominantly used for screening, as they are sufficient for comparison, see Table 6.1. These can also be advantageously implemented in automated workflows with reasonable effort. For this reason, many industrial but also manual application techniques such as brushing, rolling, electrostatic atomization, high-rotation atomization, powder coating (spraying and sintering), dip coating (conventional and electrostatic), casting, rolling and printing processes are not used in laboratory automation systems.

The implementation of the methods for drawdown and spraying as well as their variants are sometimes realized differently depending on the manufacturer. There are also major differences in terms of the substrates that can be used and their formats. One of the difficulties with contaminated application devices is cleaning them. It does not matter what shape or size a drawdown bar has. Interestingly, suppliers have been able to develop concepts for the spray process that do not require cleaning. This does contribute positive to throughput and sustainability as no cleaning media such as solvent are needed, despite low or high material turnover.

Table 6.1: Overview matrix of application methods offered for lab automation

Application method	**Chemspeed Technologies**	**Füll Lab Automation**	**Labman Automation**	**Others**
Drawdown	X	X	X	(X)[1]
Spraying	X	X	X	(X)[2]
Spin coating	X			(X)[3]

1 Semi-automated drawdown without any automated support functions requiring manual steps
2 Semi-automated lab spray stations with manual feeding and cleaning
3 Semi-automated spin coater, main application in semiconductor industry

6.1 Drawdown application

Drawdown is a key technique in paint laboratories used to evaluate the properties of paints, coatings, and related materials. It involves spreading a uniform film of paint or coating over a substrate to study its characteristics. The substrate can vary in material type and dimension. The type of substrate can be based on metals, paper, cardboard, plastic films, glass, wood, etc. Their use depends mainly on the properties to be measured. For example, the use of white-black contrast cards on cardboard is mainly intended for determining the optical properties of architectural coatings. On the other hand, industrial and automotive paints are tested on various metallic (steel, aluminium, etc.) substrates, some of which are pre-primed in the case of top coats, for optical, mechanical and chemical properties. Drawdown bars and blades come in a variety of types, each designed to achieve specific film thicknesses, finishes, and application requirements. The choice of bar or blade depends on the properties of the coating material, the desired film thickness, and the type of testing being performed. In paint and coating applications the use of wound bars is rather rarely used. The dominant types are fixed gap bars and doctor blades, but also chamber bars. Multi-gap bars are present a special use for testing the paints and coatings for their application behaviour such as sagging resistance. What they all have in common is the requirement for high, consistent, reproducible and defect-free quality.

There are now many simple film applicators on the market which already make a decisive contribution to improving the quality of drawdown applications. The substrates are usually fixed in place using a vacuum and lie flat on the application table. In addition, the application process is started by the user after the paint has been applied and the drawdown bar moves over the

Figure 6.1: Overview of "Flex Draw Down" module
Source: Chemspeed Technologies

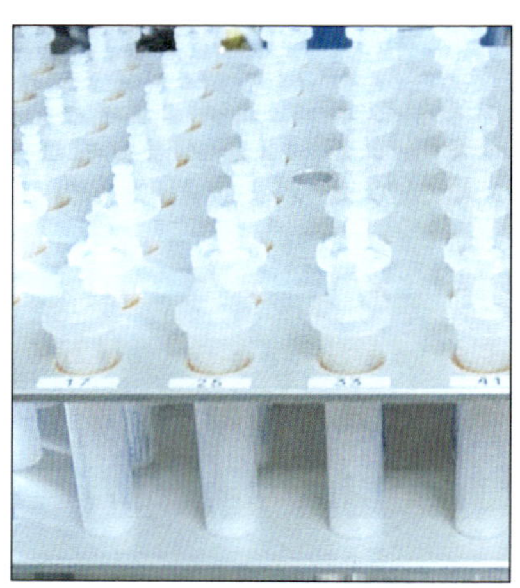

Figure 6.2: Storage of disposable syringes
Source: Chemspeed Technologies

Figure 6.3: Transfer of substrate from carousel storage to vacuum table
Source: Chemspeed Technologies

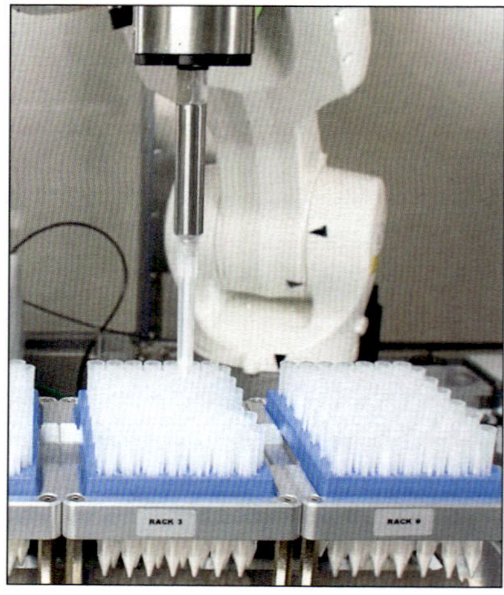

Figure 6.4: Mounted tip from storage for paint aspiration
Source: Chemspeed Technologies

substrate at a constant speed. In this respect, there are similarities between film applicators and fully automatic drawdown application. The differences are the manual processes relating to substrates and cleaning the drawdown bars.

The principle of drawdown application has been used by automation manufacturers for many years. Compared to other application methods, it is the easiest to implement. The principle is based on the use of disposable syringes, which are already pre-filled with the paint or coating material. Empty disposable syringes can also be used to aspirate the wet sample and then dispense it in front of the drawdown bar. Disposable syringes because the procurement costs are not disadvantageous in relation to the cleaning effort, but above all to avoid contamination between the tests.

The Figures 6.1 and 6.2 illustrate their use by examples of Chemspeed. In accordance with the aim of realizing experiments with a high throughput, appropriate stocks of consumables such as syringes and substrates as well as other auxiliaries are also required. These must be matched to the calculated throughput rate. The syringes to be used must of course be automation-capable so that the robots can grip and operate them.

The first step in the automated drawdown application is the provision of the substrate, which in the first example is taken from a storage position in a cassette. There is space for 10 cassettes per carousel, each with 20 substrate positions and therefore a total of 200 substrates. The carousels can also be multiplied if the capacity with one is insufficient. The substrate, with a maximum height of

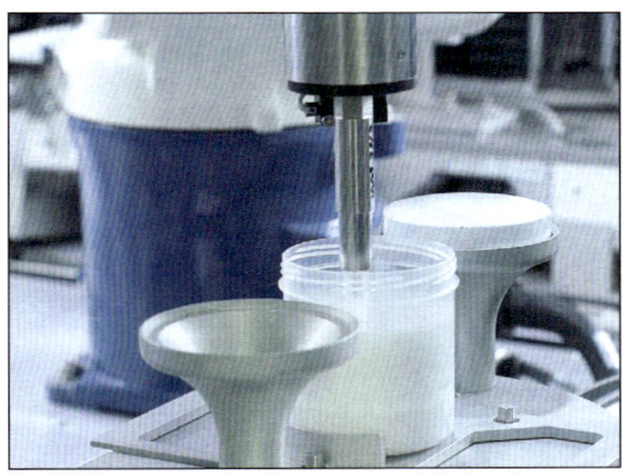

Figure 6.5: Aspiration of paint from sample vessel
Source: Chemspeed Technologies

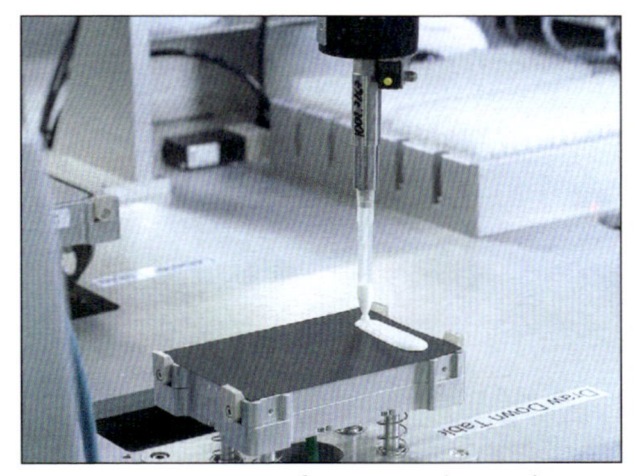

Figure 6.6: Dispensing of paint sample on substrate
Source: Chemspeed Technologies

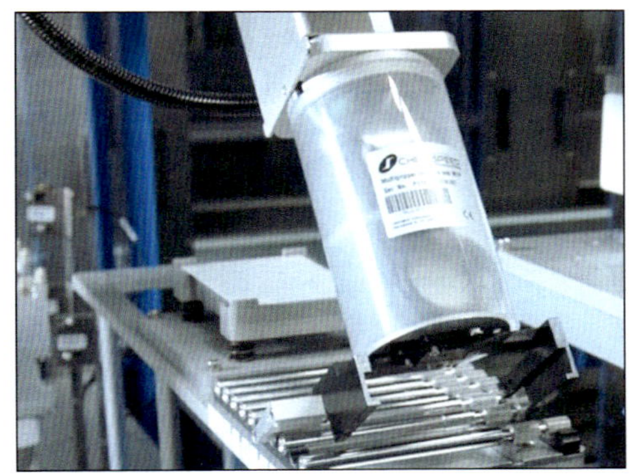

Figure 6.7: Pick up of drawdown bar from storage
Source: Chemspeed Technologies

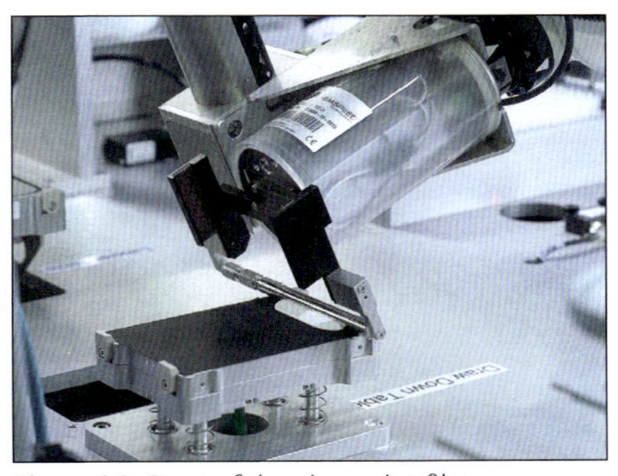

Figure 6.8: Start of drawing paint film
Source: Chemspeed Technologies

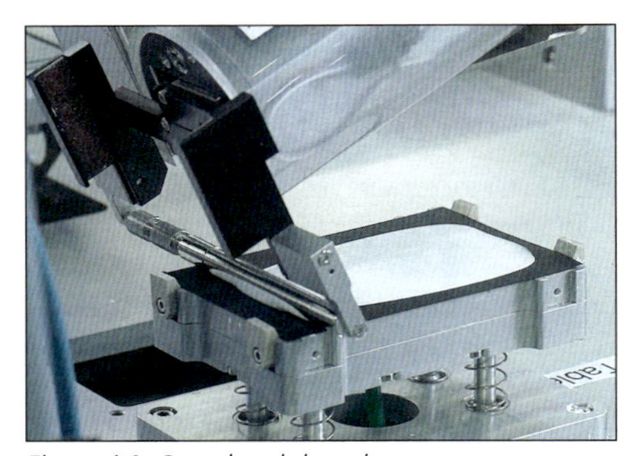

Figure 6.9: Completed drawdown
Source: Chemspeed Technologies

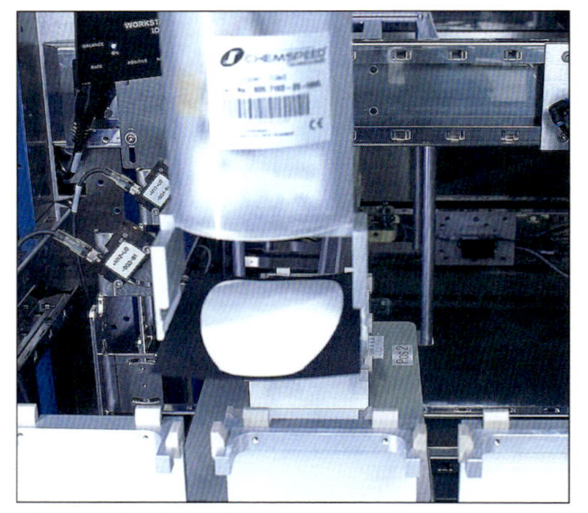

Figure 6.10: Transfer of coated substrate by multi gripper
Source: Chemspeed Technologies

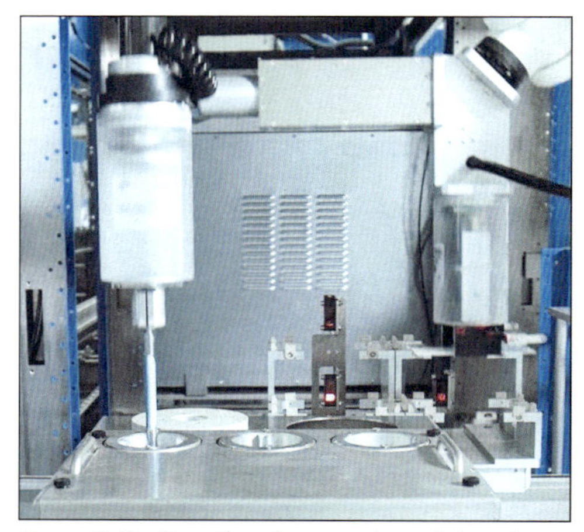

Figure 6.11: Transfer of contaminated drawdown bar into active washing station
Source: Chemspeed Technologies

5 or 10 mm depending on the configuration, is placed on a vacuum table. A tip or syringe, dependent on the selected configuration, is then retrieved from the storage using the multi-tool mounted on the multi-axis robot. After mounting, the wet sample is aspirated from the formulation vessel, which was delivered to the "Flex Draw Down" application module by means of a shuttle shown in Figure 6.3 to 6.5.

After sample aspiration, the robot moves to the drawdown table and dispenses the predefined amount of paint onto the substrate. If the volume of the tip is not sufficient, for example with larger substrate formats, this process can be repeated with the same tip. Alternatively, a syringe with several ml can be used for larger substrate formats. Once the dispensing process is complete, the tip or syringe is discarded into the waste container. The multitool on the multi axis robot picks up a drawdown bar from the store and moves to the starting position of the drawdown, see Figures 6.6 to 6.8.

The multi-axis robot applies the paint with uniform precision and repeatability. The smallest differences in the robot movement can also be cushioned by the damping mechanism of the vacuum table. The paint-contaminated drawdown bar is then transferred to the active washing station. There, depending on the setup, an overhead robot takes over the bar and cleans it in several washing chambers and then

Figure 6.12: Overview of "Flex Draw Down" with linear axis application and frame bar
Source: Chemspeed Technologies

dries it before transferring it back to the storage area. In the meantime, the coated substrate is transferred by the gripper fingers of the robot to the next workflow step, which is not always, but usually, drying, shown in Figure 6.9 to 6.11.

Depending on user requirements, Chemspeed offers further drawdown application solutions. In the following example, a raw material sample from production is compared with a reference sample for quality control purposes. For this purpose, the two coatings were produced in the formulation module and now applied to contrast cards in the drawdown module. A parallel drawdown is implemented for better, also visual, comparability. For this purpose, a larger format of the substrate is placed on the vacuum table followed by the double chambered drawdown applicator, Figure 6.12 and 6.13. Afterwards, the two formulation vessels are transferred from the shuttle to the drawdown station using a multi-axis robot.

The special feature of this example is the filling process of the double chambered drawdown applicator. First the cup with the coating batch to be checked and then the cup with the reference sample is poured into one of the chambers. The parallel drawdown process then starts. In contrast to the previous example, this time the double chambered drawdown applicator is moved evenly and precisely over the substrate by a linear axis, illustrated in Figure 6.14 to 6.16.

Figure 6.14: Pouring sample 1 from formulation vessel into frame 1 Source: Chemspeed Technologies

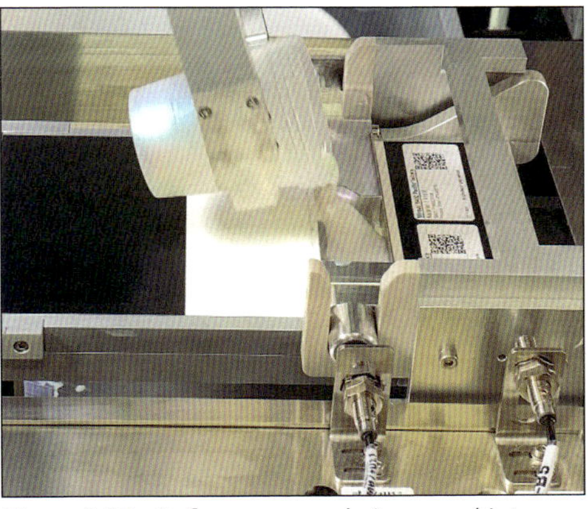

Figure 6.15: Reference sample is poured into frame 2 Source: Chemspeed Technologies

Figure 6.13: Transfer of open paint vessel from shuttle to drawdown station Source: Chemspeed Technologies

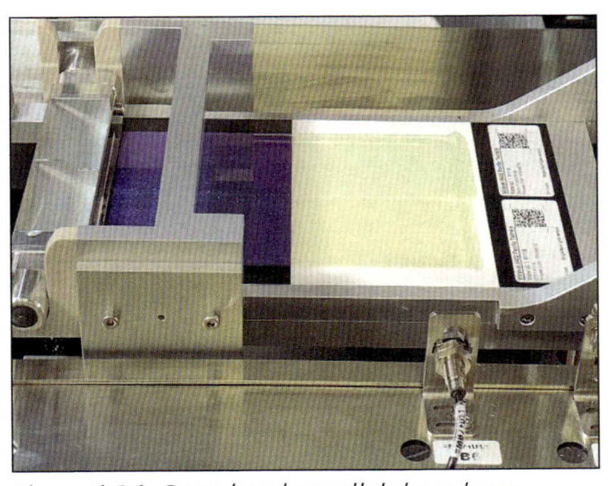

Figure 6.16: Completed parallel drawdown Source: Chemspeed Technologies

The coated contrast card is lifted from the vacuum table using vacuum buttons on the edge and placed on a waiting shuttle for the next workflow step. The multi axis robot then picks up the drawdown applicator with its multi tool and transfers it to the cleaning station. The cleaning station is then closed, and the applicator is thoroughly cleaned using solvent from the storage drum. In the meantime, the robot picks up the next contrast card from the card dispenser for the next test and places it on the vacuum table. Once cleaning is complete, the drawdown applicator is transferred from the cleaning station back to the drawdown applicator storage area for the next trials, shown in Figure 6.17 to 6.19.

Füll Lab Automation also relies on its "BLS syringe" concept for drawdown applications. The paint and coating samples have been produced in the formulation cartridge. These are then con-

Figure 6.17: Transfer of coated cardboard via vacuum nobs Source: Chemspeed Technologies

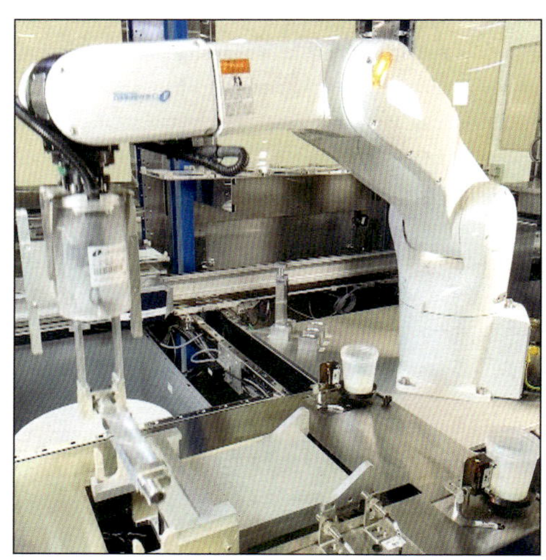

Figure 6.18: Pick up of drawdown bar for cleaning Source: Chemspeed Technologies

Figure 6.19: Cleaned frame bar is picked up from active washing station Source: Chemspeed Technologies

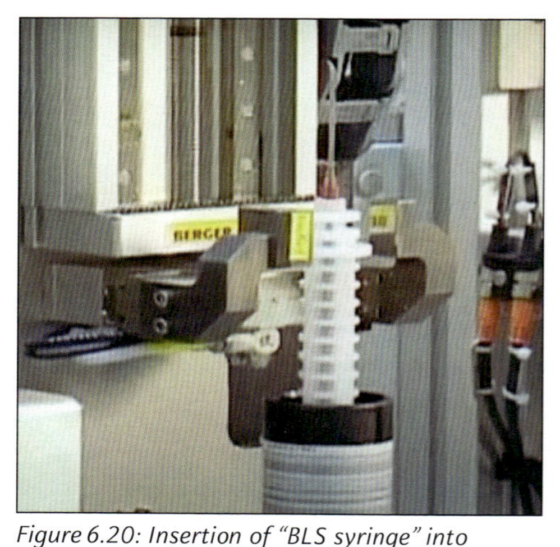

Figure 6.20: Insertion of "BLS syringe" into formulation cartridge for sample transfer Source: Füll Lab Automation

nected with the "BLS syringe" by inserting the plunger. The end of the insertion is checked by means of a sensor. The dispensing nozzle is already pre-assembled at the other end of the "BLS syringe", from which the dispensing process then takes place in the next step. In preparation for the drawdown process, a metal clamping board is retrieved from the substrate store, on which the contrast card has already been manually pre-assembled. The clamping board is placed in the intended drawdown position, Figure 6.20 and 6.21.

The next step is to load the drawdown station with one or, in the case of a parallel drawdown, with two cartridges with the inserted "BLS syringe". The dispensing nozzles of the "BLS syringes" are aligned above the single frame drawdown applicator, Figure 6.22 and 6.23. The simultaneous and exact volumetric dosing of the colours takes place side by side in the frame of the applicator. The used cartridges with the "BLS syringe" can be used again for more drawdown applications if required as the cartridge contains sufficient coating material.

The single frame film applicator is moved over the contrast card at similar speed. The weight of the film applicator seems to be sufficient to keep the contrast card flat on the clamping board without vacuum support. Both applied paints stay separated and do not show flow effects into each other. When finished, the single frame film applicator is transferred to the cleaning bath in the washing station, just behind the drawdown station for cleaning purposes. The completely

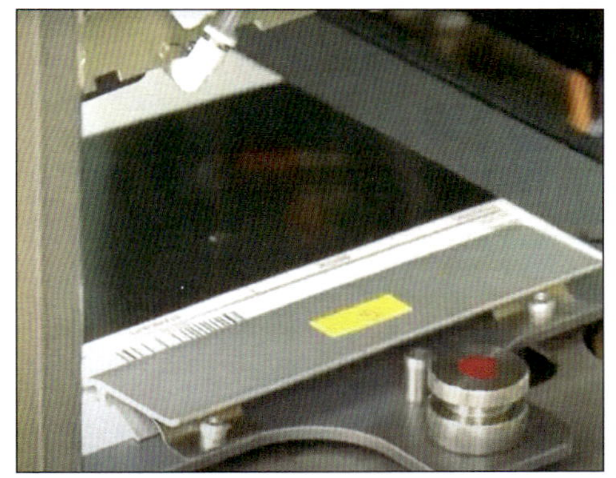

Figure 6.21: Positioned clamping board with pre-loaded contrast card in drawdown station
Source: Füll Lab Automation

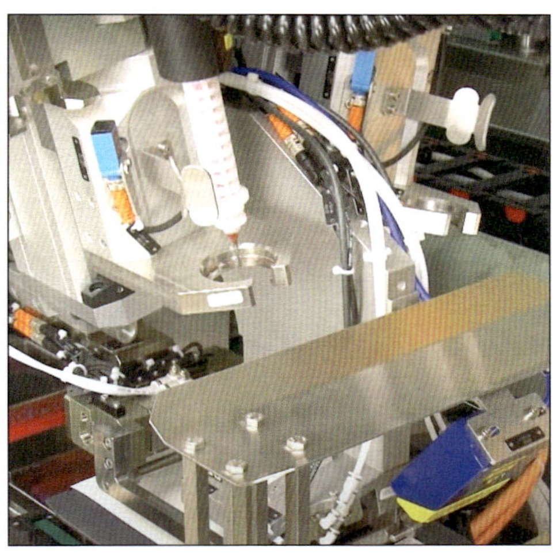

Figure 6.22: Insertion of cartridge with "BLS syringes" into drawdown dispensing station
Source: Füll Lab Automation

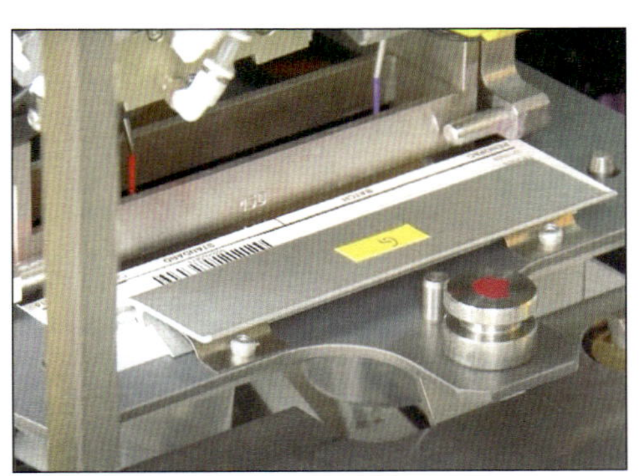

Figure 6.23: Dispensing of two paint samples for parallel drawdown
Source: Füll Lab Automation

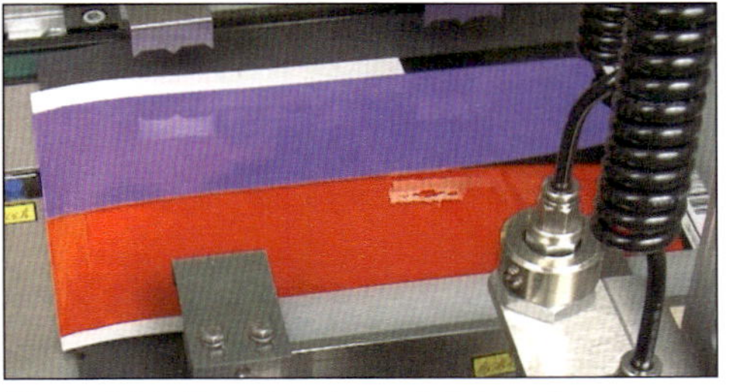

Figure 6.24: Completed parallel drawdown
Source: Füll Lab Automation

cleaned film applicator is afterwards dried by means of an air flow provided via nozzles. The dried film applicator is then ready to be used for the next drawdown, see Figure 6.24 to 6.26.

The next presented example is from Labman and called "Automated Drawdown System". The overview presents a well cleaned platform consisting of a syringe storage, drawdown tables, drawdown bar storage and substrate storage as well as a "hidden" semi-automated solvent cleaning bath below the platform, Figure 6.27 and 6.28.

The system dispenses sample from a syringe onto the desired surface or mould, consisting of an "XYZ gantry" and stacker axis to receive drawn down panels. The process starts with the system

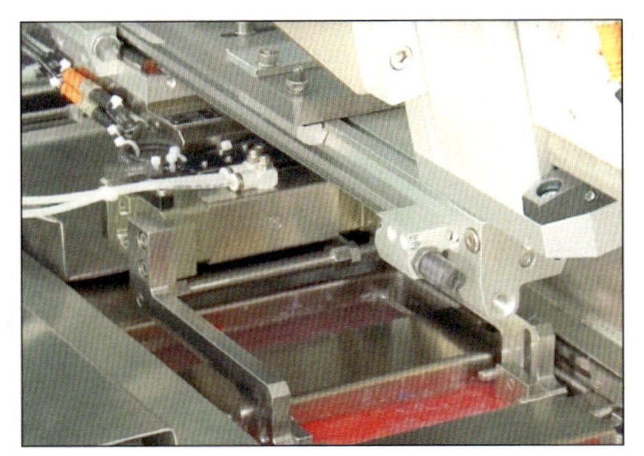

Figure 6.25: Transfer of drawdown bar into cleaning station Source: Füll Lab Automation

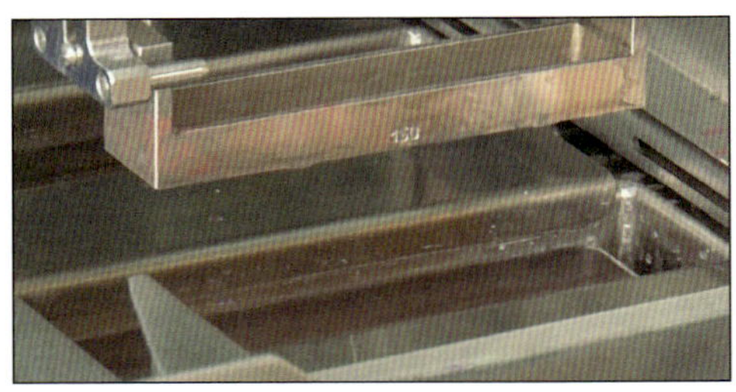

Figure 6.26: Pick up of cleaned drawdown bar Source: Füll Lab Automation

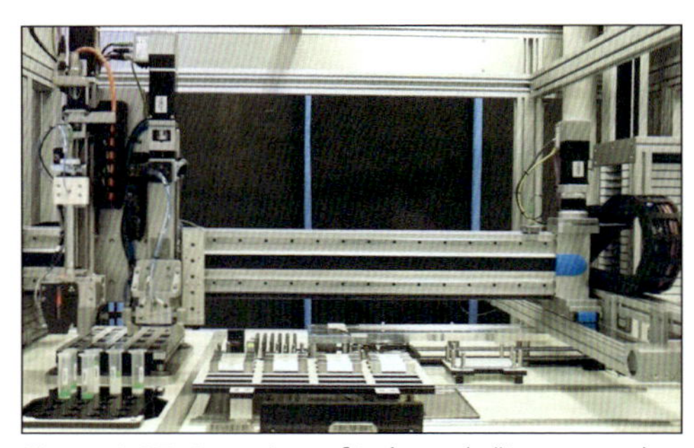

Figure 6.27: Overview of Labman's "Automated Drawdown System" Source: Labman Automation

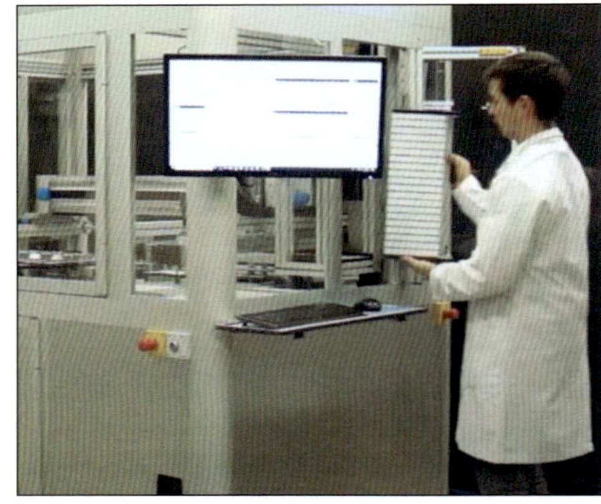

Figure 6.28: Loading of substrate tray into storage Source: Labman Automation

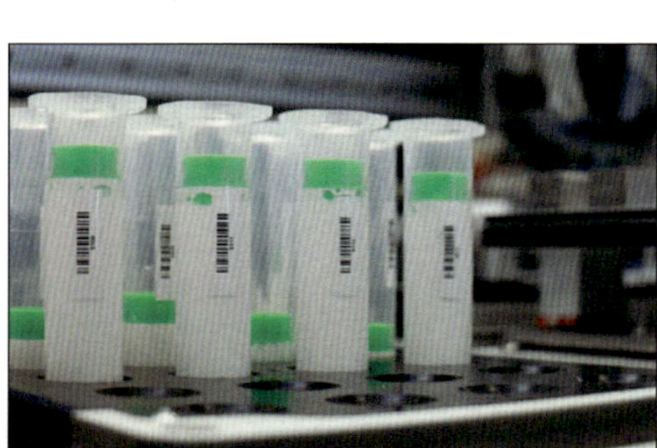

Figure 6.29: Loaded storage of pre-filled syringes with sample material Source: Labman Automation

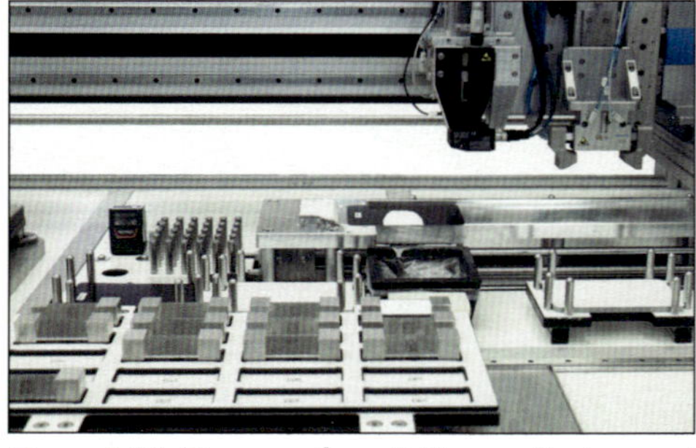

Figure 6.30: Storage of small film applicators and drawdown bars Source: Labman Automation

picking a syringe, reading the barcode and de-capping it. For slide drawdown, a die is placed on the syringe tip. The system dispenses a measured amount of sample and draws downs using the die onto 20 glass slides secured in a universal slide rack, Figure 6.29 and 6.30.

A mould rack can replace the slide rack, in which case sample is dispensed without a die into the desired moulds. The system provides compatibility for the use of many different moulds, panels and slides. Intuitive software allows the user to control many features in the process, changing run type, consumable type and dispense characteristics. Supported is this by automatic capping and de-capping of syringes, see Figure 6.31 to 6.33.

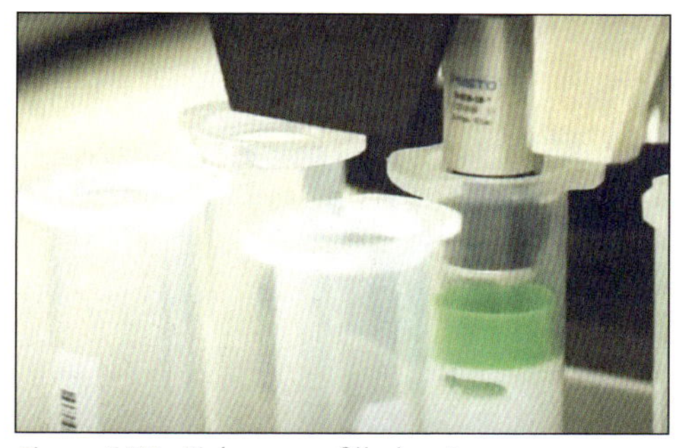

Figure 6.31: Pick up pre-filled syringe
Source: Labman Automation

Figure 6.32: Mounting small film applicator to syringe
Source: Labman Automation

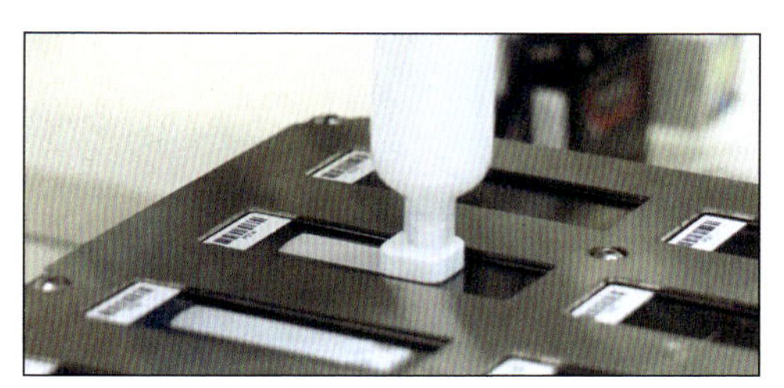

Figure 6.33: Drawdown into mould
Source: Labman Automation

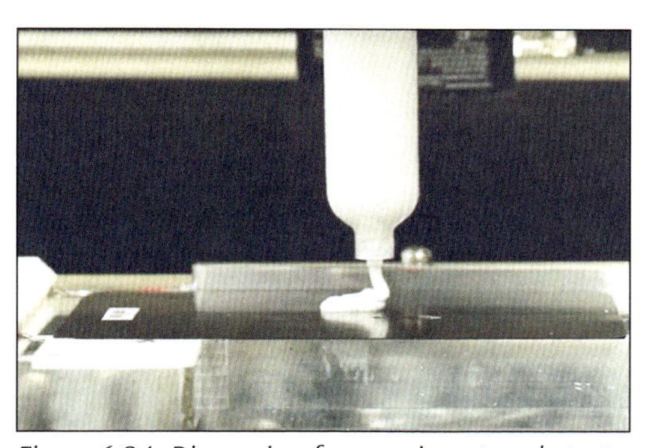

Figure 6.34: Dispensing from syringe to substrate
Source: Labman Automation

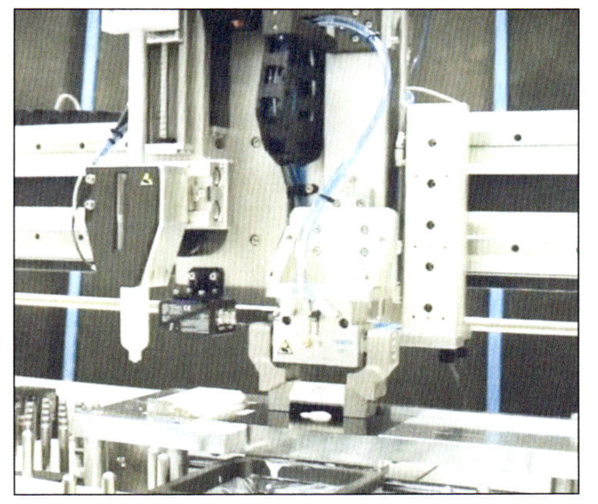

Figure 6.35: Drawdown of paint sample
Source: Labman Automation

Figure 6.36: Cleaning bath with draw down bar
Source: Labman Automation

For panel drawdown, a panel is placed on the vacuum bed where the syringe dispenses a measured amount of sample. The paint samples are applied with 4 sided "TQC bird bar" applicators, allowing for a range in coating thickness. The bird bar, picked up from the storage, is moved across the panel creating a thin film of sample. The custom vacuum bed keeps panels flat during the drawdown. At the end the panel is pushed into the stacker and the process is repeated, Figure 6.34 to 6.36.

For the purpose of drawdown applications, Labman offers second standalone concept. The "Panel Preparation System" is loaded with formulation vessels of paint and blank panels. A blank panel is picked from the panel stack and loaded onto a vacuum bed which holds the substrate flat to ensure accurate and repeatable drawdowns, see Figure 6.37 and 6.38.

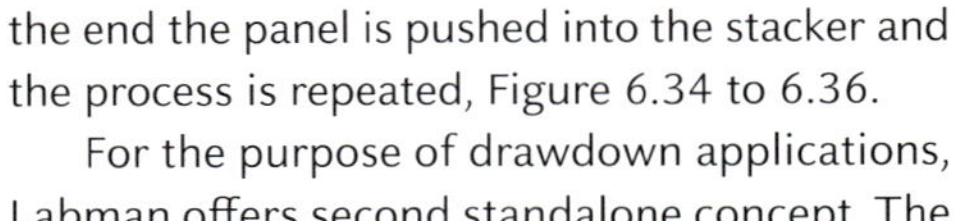

Figure 6.37: "Panel Preparation System" of Labman
Source: Labman Automation

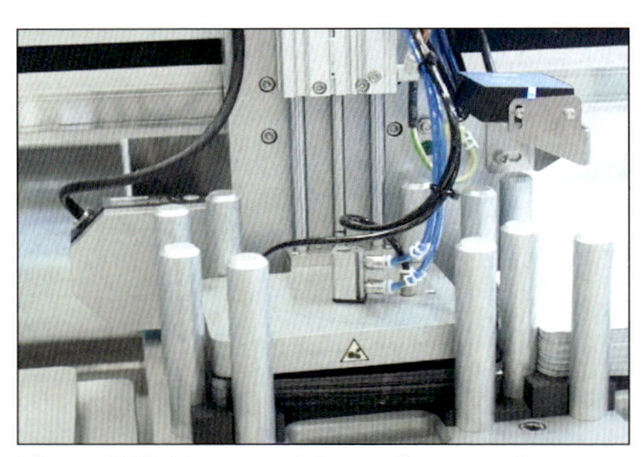

Figure 6.38: Vacuum pick up of uncoated substrate from storage
Source: Labman Automation

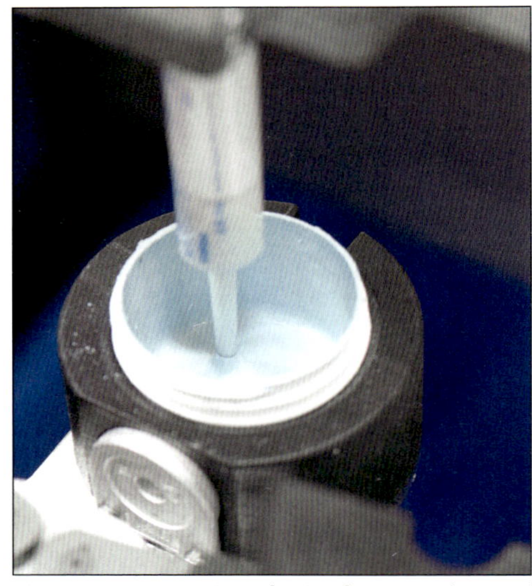

Figure 6.40: Aspiration of paint from sample vessel
Source: Labman Automation

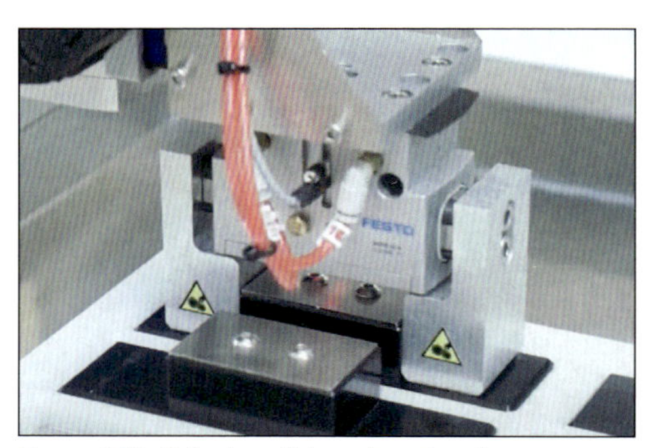

Figure 6.39: Gripping of drawdown bar from storage
Source: Labman Automation

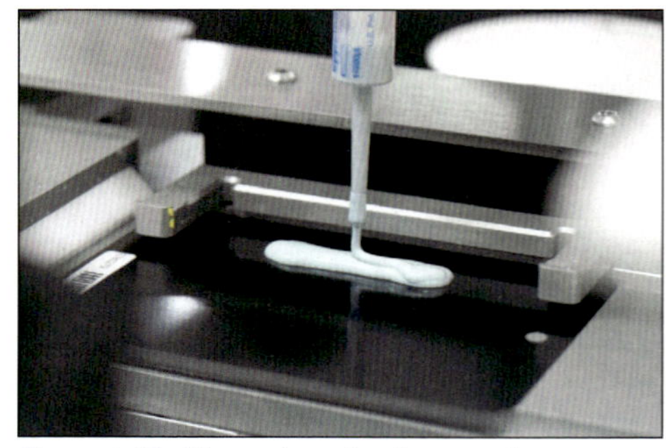

Figure 6.41: Dispensing of paint sample onto substrate
Source: Labman Automation

Each formulation vessel is picked from its rack location, has its barcode read and is then mixed in a dual asymmetric centrifuge (DAC) and de-capped. For the aspiration from the vessel and dispensing of paint onto the substrate in a line, 10 ml disposable syringes are used from the storage rack. A drawdown bar is then pulled across the substrate to create a thin film, see Figure 6.39 to 6.41.

Afterwards the drawdown bar is washed and dried. The washing and drying of drawdown bars between samples prevents cross contamination. Finally, the substrate with the wet coating film is picked up from the vacuum bed and stored in a removable cassette to dry, see Figure 6.42 and 6.43.

6.2 Spray application

Automated spraying has been used for decades to apply mass-produced coatings to car bodies and other objects. The difference in production is the less frequent material changes and the larger quantities of material that are processed. In the development of paints and coatings, however, the constant change of samples makes automation difficult and requires a lot of cleaning.

Figure 6.42: Coated substrate on drawdown table
Source: Labman Automation

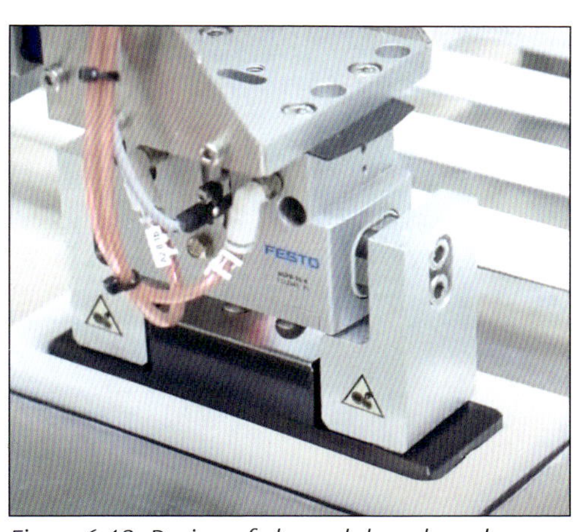

Figure 6.43: Drying of cleaned drawdown bar
Source: Labman Automation

Figure 6.44: "BLS Spray Application"
Source: Füll Lab Automation

Figure 6.45: Placing of substrate into spray cabin
Source: Füll Lab Automation

The primary goal is to avoid contamination of individual paint samples, whether due to a different colour or formulation composition. The first company which developed a system for contamination-free spray application despite zero cleaning in lab environment was the former Bosch Lab Systems. The innovative technology was winning the European Coating Award in 2005. The business of Bosch Lab Systems has been transferred into the new founded company Füll Lab Automation in 2021 [5]. The "BLS Spray Application" allows highly reproducible coating of test panels in industrial quality with low amounts of paint and switching between colours without cleaning for water- and solvent based paints and coatings, see Figure 6.44 and 6.45.

After loading a formulation cartridge with mounted "BLS syringe" with the coating material (water or solvent based) and the test panel into the spray zone, selecting the spraying parameters, the spraying process is carried out fully automatically, see Figure 6.46 and 6.47. The innovative design is based on a conventional pneumatic spray head with up to 10 bar ensuring highly reproducible spraying results. The award-winning technology makes it possible to switch between different paints and coatings without cleaning. Different pneumatic spray heads and nozzle types are available.

Figure 6.46: Tool change from gripper to syringe function Source: Füll Lab Automation

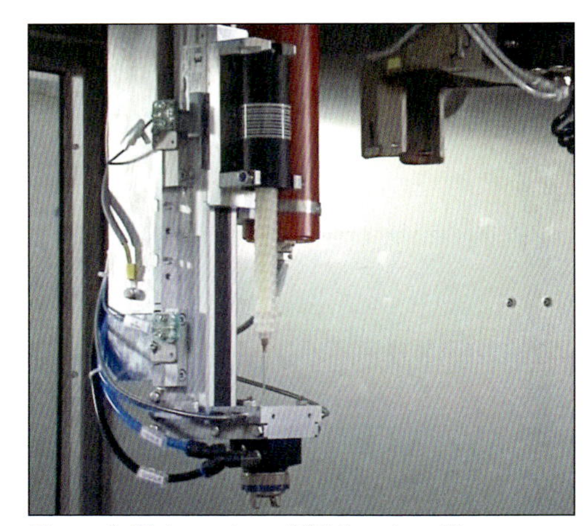

Figure 6.47: Inserting of "BLS syringe" into spray head Source: Füll Lab Automation

Figure 6.48: Automated spraying on horizontal positioned substrate Source: Füll Lab Automation

Figure 6.49: Pick up of substrate after spraying Source: Füll Lab Automation

The "BLS Spray Application" is available in different sizes and different variants, see Figure 6.48 to 6.49. Depending on the requirements a small stand-alone machine or a completely automated system might be ideal for the use of a R&D coating-lab or even in QC environment. The "BLS Spray Application" can also be combined with Füll Lab's "Compact Lab Station" or integrated into their "Integrated Lab Station".

The coating results are independent from the operator respectively from the human factor. Depending on the system configuration it allows the spraying of up to 400 test panels per day. Already small amounts of paint (≥ 5ml), depending on the required coating thickness and substrate size, are sufficient to receive homogeneous coating films. The system can come with different configurations, e.g. for horizontal or vertical spraying. Füll Lab Automation also offers rotary bell application, electrostatic spraying or airless application with their automated spray system.

The second manufacturer of fully automated spraying systems, Chemspeed Technologies, has also developed a cleaning-free spraying concept and had it patented. In contrast to Füll Lab, the paint is not fed through the spray head from the rear, but from the side of the spray head in front

Figure 6.50: Vertical mounting of substrate
Source: Chemspeed Technologies

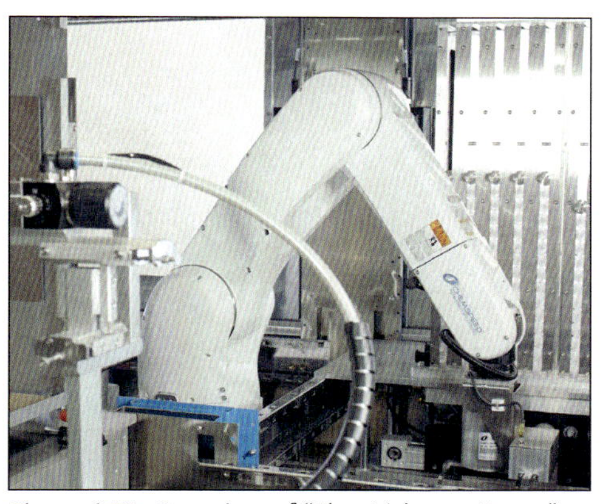

Figure 6.51: Overview of "Flex Airborne Spray"
Source: Chemspeed Technologies

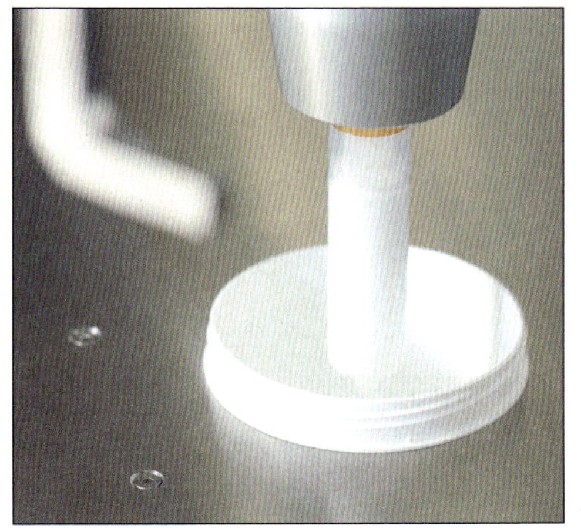

Figure 6.52: Aspiration with 20 ml standard syringe
Source: Chemspeed Technologies

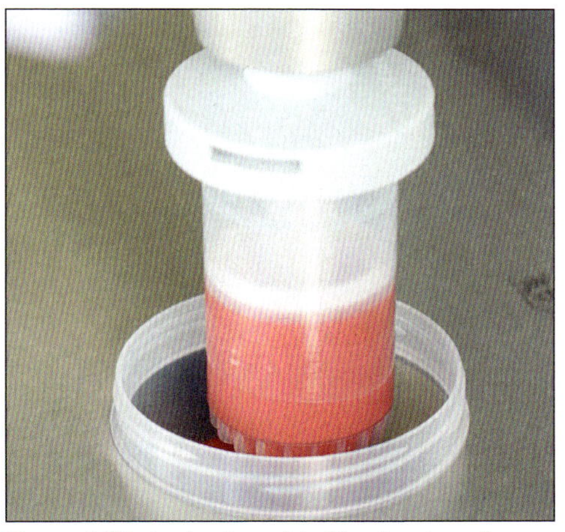

Figure 6.53: Aspiration with 50 ml syringe for larger panels
Source: Chemspeed Technologies

of the atomizer nozzle. This means that the second patented process also allows painting without cleaning. This avoids the use of ecologically harmful solvents for cleaning. In other words, the use of laboratory automation also has a positive impact on sustainability and reducing the carbon footprint. The systems can be operated with water- and solvent-based as well as 1-pack and 2-pack coating systems, see Figure 6.50 to 6.51.

Chemspeed offers the concept in two versions. The first system presented is offered as a compact spray module integrated into a "Flexshuttle" or as a standalone solution. To load the system, the substrate is first fixed vertically onto the sample holder by vacuum nobs from the side, which is opened for this purpose and closed again during operation. The sample holder then turns 90 degrees to the start position.

Next, the sample vessel is opened using a screw capper before the robot mounts a disposable, commercially available syringe. This is then used to aspirate the paint from the sample container. Typically, syringes with a volume of 10 or 20 ml are used, depending on the substrate size and target layer thickness, see Figure 6.52. When coating several or larger substrates or multiple coating layers, the system can also be configured with even larger syringes of 50 ml, see Figure 6.53.

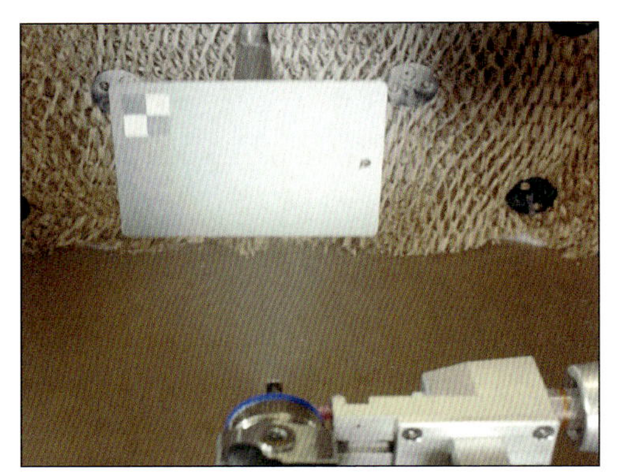

Figure 6.54: Vertical spraying of standard substrate size Source: Chemspeed Technologies

Before the robot positions the loaded syringe in front of the nozzle head, a disposable tip with a defined nozzle size is mounted onto the syringe. The exact positioning of the tip in front of the nozzle head is realized by a corresponding tip guide. The centre and horn air are used to atomize the paint very finely and adjust the spray image. The amount of paint applied is controlled by the finely adjustable piston stroke of the syringe, see Figure 6.54. As usual when spraying, the substrate can be rotated 90 degrees by the multi-axis robot. This makes it possible to achieve real spray cross-coats for optimum overlapping on the substrate, see Figure 6.55 and 6.56.

After the coating process, the substrate holder turns to the side again and the substrate is re-

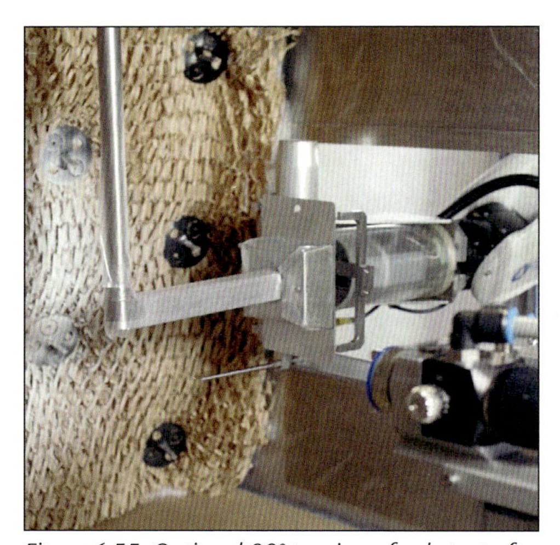

Figure 6.55: Optional 90° turning of substrate for next spray layer Source: Chemspeed Technologies

Figure 6.56: Turned substrate is sprayed with next layer Source: Chemspeed Technologies

moved from the robot and placed in a parking position to flash off. Alternatively, the substrate can be placed directly on a shuttle and transported to the next workflow step. This could, for example, be drying at room temperature or heat curing in an oven.

Many parameters can be adjusted when applying paints and coatings using Chemspeed's "Flex Spray" similar to the "BLS Spray Application", e.g. individual pressure adjustment of atomization and horn air, sample dispensing speed, tip nozzle size, spraying distance, spray overlap, axis speed, spraying pattern and various setups are available to allow working with a range of substrate sizes up to DIN A4, see Figure 6.57 and 6.58. In addition, the complete spray head can be swivelled by 90 degree and combined with a different shaped spray holder, resulting in horizontal spraying of the substrate

If spraying of multiple substrates at the time is required, then the large "Flex Spray Airborne L" automation solution can be obtained, see Figure 6.59 and 6.60. The fully automated paint application is also qualified for large panel formats or up to 4 standard substrate sizes, see Figure 6.61. The large version of airborne spray is only available as integrated module into a "Flexshuttle" Paints & Coatings or "Flex-

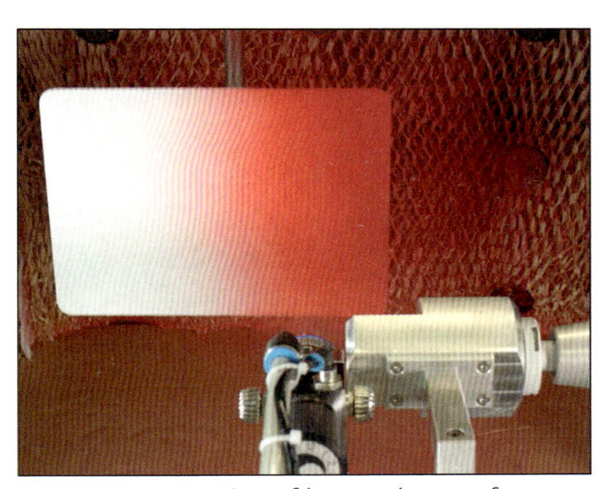

Figure 6.57: Spraying of large substrate format Source: Chemspeed Technologies

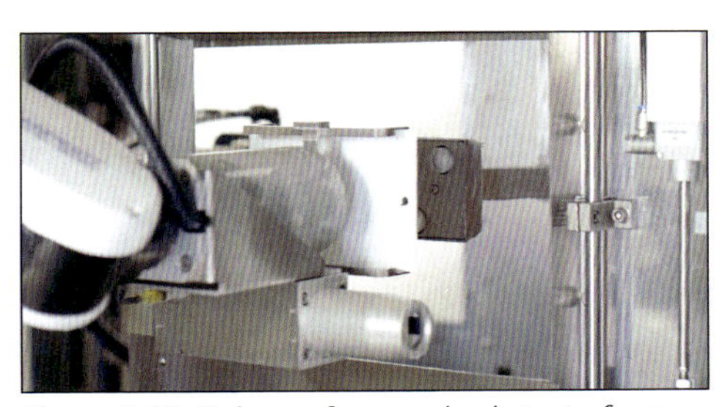

Figure 6.58: Pick up of sprayed substrate from spray cabin Source: Chemspeed Technologies

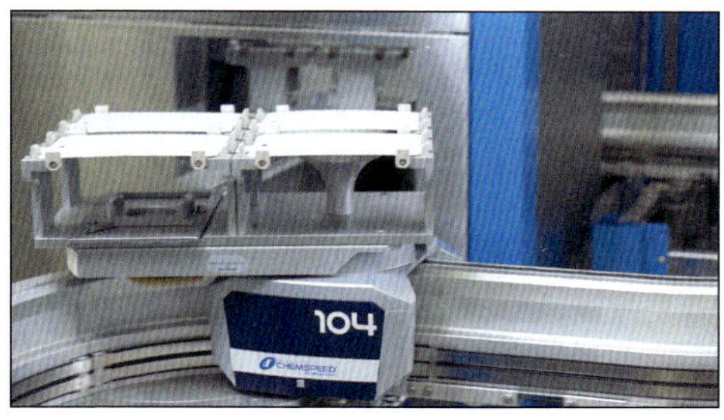

Figure 6.59: Shuttle with 4 substrate arriving at "Flex Spray Airborne L" Source: Chemspeed Technologies

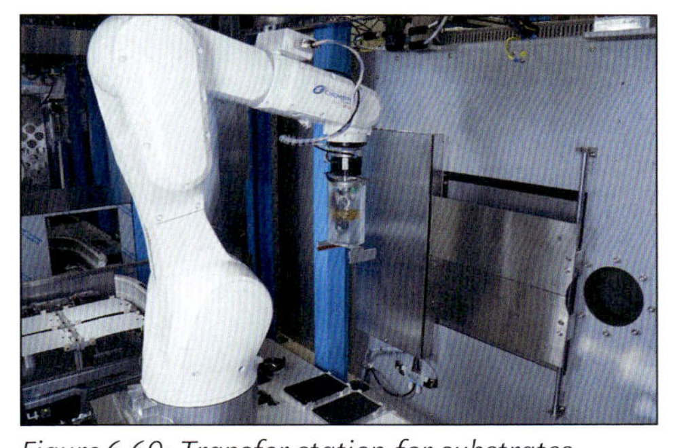

Figure 6.60: Transfer station for substrates Source: Chemspeed Technologies

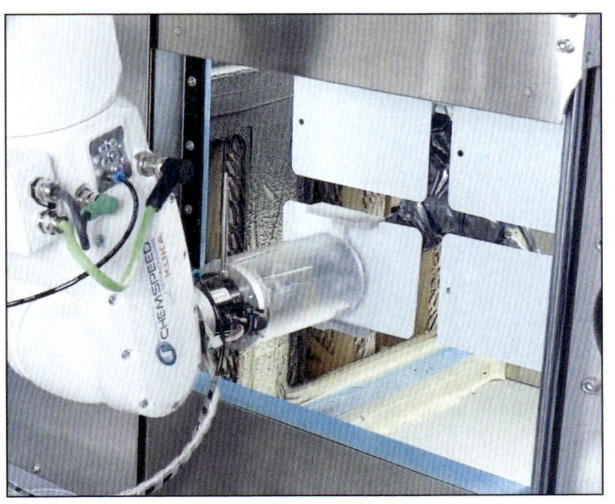

Figure 6.61: Mounting of 4 substrates onto vacuum holder Source: Chemspeed Technologies

shuttle Colmatch". The "Flex Spray Airborne L" also allows spraying on vertical or horizontal positioned substrates.

The workflow begins with the arrival of a shuttle in the "Flex Spray Airborne L", which can carry up to 4 substrates. These are unloaded by the centrally positioned multi-axis robot and mount-

Figure 6.62: Tool change from substrate gripper to syringe *Source: Chemspeed Technologies*

Figure 6.63: Level detection of filling level in vessel *Source: Chemspeed Technologies*

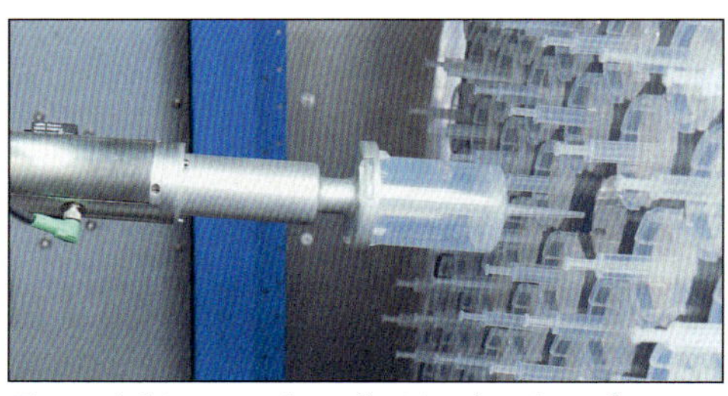

Figure 6.64: Mounting of 100 ml syringe from storage *Source: Chemspeed Technologies*

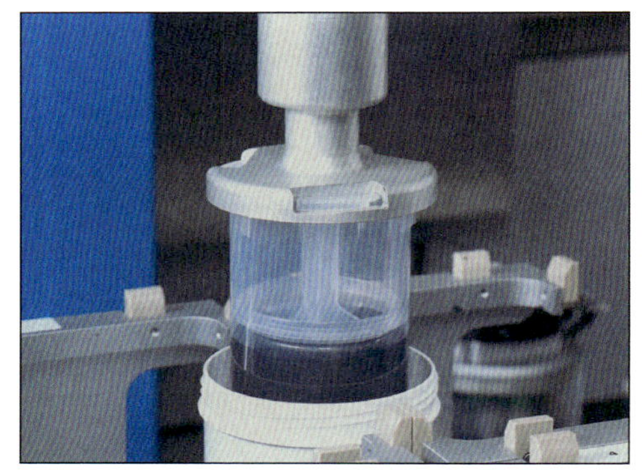

Figure 6.65: Aspiration of coatings sample *Source: Chemspeed Technologies*

Figure 6.66: Mounting of spray tip *Source: Chemspeed Technologies*

Figure 6.67: Spray tip is positioned from the side in front of the spray *Source: Chemspeed Technologies*

ed one after the other with the gripper fingers on the substrate holder in the spray booth and fixed in place using a vacuum. If fewer substrates are to be coated, appropriate blind substrates must be loaded to protect the vacuum nobs. Alternatively, large-format substrates can also be mounted. As with the small spray setup, the substrate holder swivels through 90° and the side doors close.

After loading the substrates, the substrate tool head of the robot is replaced with the syringe tool head, see Figure 6.62. The next step in the workflow is to measure the fill level of the paint in the vessel using a sensor. Depending on the formulation, the fill level may vary, and this prevents the syringe from being immersed too deeply in the paint sample, see Figure 6.63. This prevents contamination of the system during the subsequent transport processes and the tip for spraying can be picked up correctly. However, the robot first retrieves a 100 ml disposable syringe from the storage panel, which is large enough to paint 4 standard formats or a large substrate with even multiple layers of paint, see Figure 6.64.

During aspiration of the paint, the syringe automatically dips deeper into the cup due to the decreasing fill level, see Figure 6.65. This prevents the syringe from drawing up air. After aspiration, the tip with the defined nozzle size is picked up on the syringe tip, see Figure 6.66. Now the robot inserts the complete spray arm into the shielded channel, which is necessary to ensure the ATEX design of the spray booth. The tip is positioned exactly in front of the spray head by the tip guide so that a perfect spray pattern is achieved, see Figure 6.67.

As with its smaller sister system, the substrate holder moves in the X-Y direction during the coating process. It is possible to define various parameters such as intermediate flash-off and the application of several layers wet-in-wet. Similar to the "BLS Spray Application", the system can also be configured with a "Spray Spy"[6]. This enables simultaneous measurement of the speed and size of droplets, droplet impulse and volume flow as well as separate evaluation of transparent and non-transparent droplets, etc. for qualitative and quantitative analysis of the spray process. After the spray process, the spray arm is moved out of the shielded channel and the spray tip and syringe are disposed of after optional emptying, see Figure 6.68 to 6.70.

After the tool change, the coated substrates are temporarily stored for intermediate flash-off or loaded directly back onto the shuttle for the next workflow step, for example heat curing.

Figure 6.68: Spraying of 4 substrate at the same time in ATEX spray cabin

Source: Chemspeed Technologies

Figure 6.69: Withdraw of syringe from spray cabin

Source: Chemspeed Technologies

Figure 6.70: Discarding of tip

Source: Chemspeed Technologies

Depending on the configuration of the "Flex Spray Systems", throughput rates of several hundred substrates per day can be achieved, see Figure 6.71 and 6.72.

There are also simpler concepts for the automation of spray processes on the supplier market. Here, however, it is no longer possible to speak of full automation, as certain manual interventions are required for operation. For example, the German manufacturer Oerter Applikationstechnik offers semi-automated spraying machines in various sizes. This requires manual

Figure 6.71: Pick up of coated substrates back to shuttle Source: Chemspeed Technologies

Figure 6.72: Loaded shuttle leaves the spray module while the next shuttle arrives Source: Chemspeed Technologies

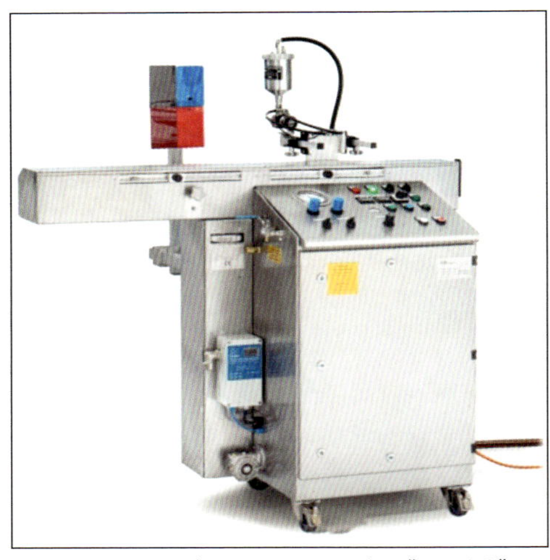

Figure 6.73: Semi-spray automation "APL 1.2" Source: Oerter Applikationstechnik

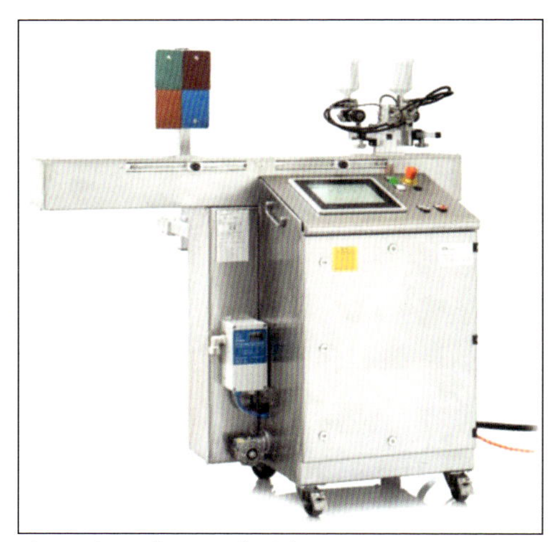

Figure 6.74: "APL 4.7" with two spray guns Source: Oerter Applikationstechnik

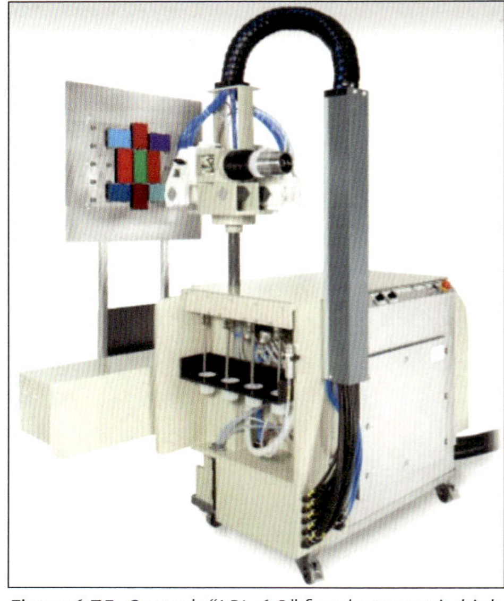

Figure 6.75: Oerter's "APL 6.3" for electrostatic high rotation bell coating Source: Oerter Applikationstechnik

loading of the paint sample and the substrates to be painted. The automatic spray systems are available with one or two flow cup guns, flexibly adjustable parameters for gun stroke speed, horizontal painting area, atomizing air control, wet-in-wet spraying, coating programs such as coating a layer thickness wedge, different substrate formats and also in an ATEX version, see Figure 673 to 6.75. After coating, the flow cup gun must be cleaned manually or at least partially for the larger systems, so that the next sample can be applied without contamination. When using the semi-automated spray systems, they must be installed in coating cabins with appropriate supply and exhaust air and overspray filtration, which is already integrated in the fully automated systems from Füll Lab and Chemspeed.

The range of equipment and the various setups are nevertheless diverse. It should be emphasized that in addition to the classic pneumatic flow cup guns, electrostatic high-rotation bell application processes are also possible. Here, spray heads from the manufacturer Dürr are used. The options include up to 4 material feed units with dosing pumps and colour changers, high-precision positioning and dosing using synchronous servomotors, painting programs, individually adjustable flash-off times, automatic quick rinsing system for pumps/paint lines, rotating test plate holder (90° rotation) for crosswise coating and semi-automatic quick-change system for changing the atomizer, see Figure 6.76 and 6.77.

6.3 Spin coating application

Spin coating is a widely used technique for applying uniform thin films to flat substrates. The process involves depositing a liquid solution onto a substrate and then spinning the substrate at high speed (typically 500 to 8000 rpm) to spread the material uniformly through centrifugal force. Typical areas of application are more likely to be found in the fields of microelectronics with application of photoresists in semiconductor fabrication and research in thin film radiation curing on silicon wafers, optics for thin films for anti-reflective or high-reflective coatings, nanotechnology, solar cells for fabrication of thin layers in organic and perovskite solar cells and biotechnologies. With spin coating, uniform film thicknesses can be achieved, and it is also a cost-efficient process with scalability for a wide range of materials.

Anyone interested in a fully automated spin coating process will currently find what they are looking for at Chemspeed. The "Flex Spin Coating" contains all the necessary components. The workflow starts with the removal of

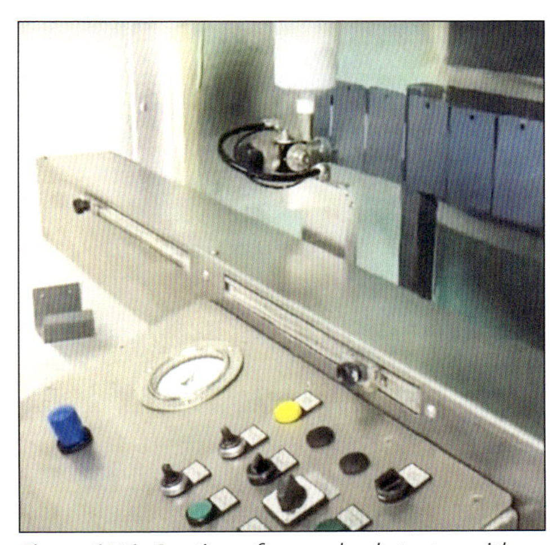

Figure 6.76: Coating of several substrates with airborne spray Source: Oerter Applikationstechnik

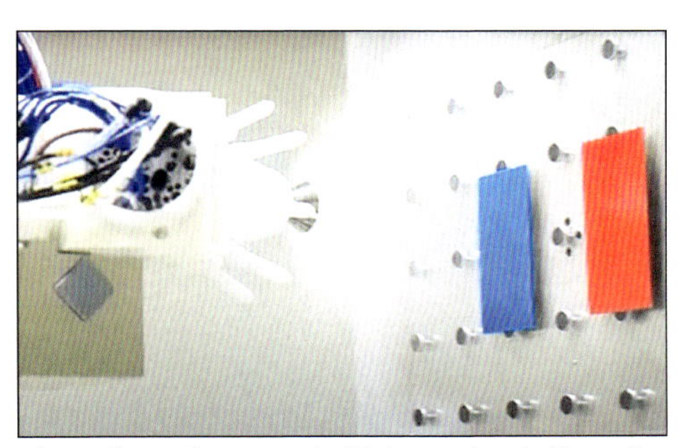

Figure 6.77: Electrostatic high rotation bell coating Source: Oerter Applikationstechnik

the substrate from the dispenser. The multi-axis robot places the substrate in the spin coater with vacuum support, see Figure 6.78 and 6.79.

Next, the robot's multifunctional tool retrieves a tip from the storage and mounts it. The coating material is then aspirated from the formulation vessel delivered via shuttle, see Figure 6.80 to 6.82. At a low rotation speed, a sufficient quantity of the coating material is dispensed onto the substrate. The inner area of the substrate remains uncoated for the further automated handling.

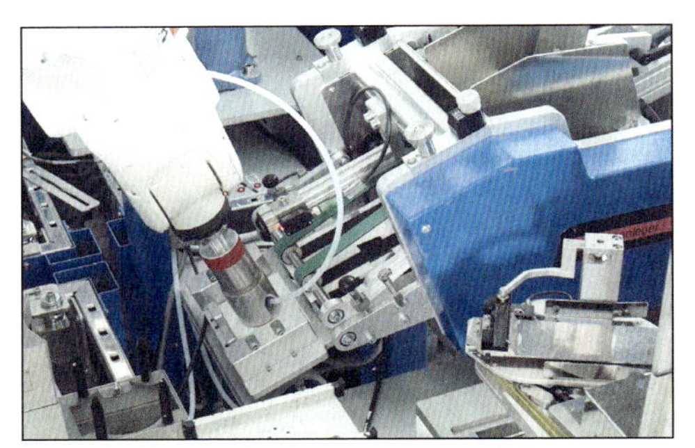

Figure 6.78: Robot picks up substrate from storage
Source: Chemspeed Technologies

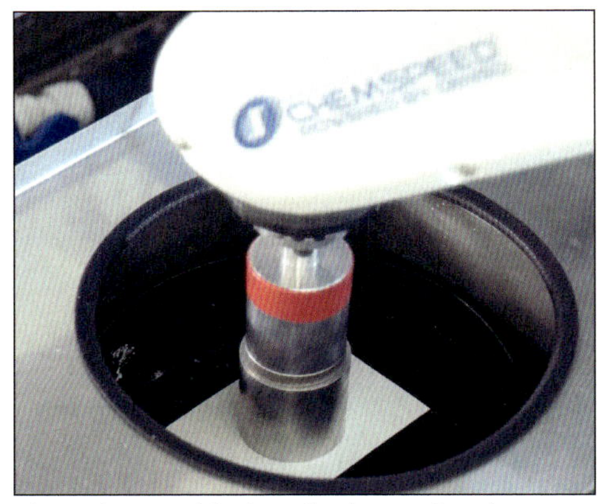

Figure 6.79: Positioning of substrate in spin coater
Source: Chemspeed Technologies

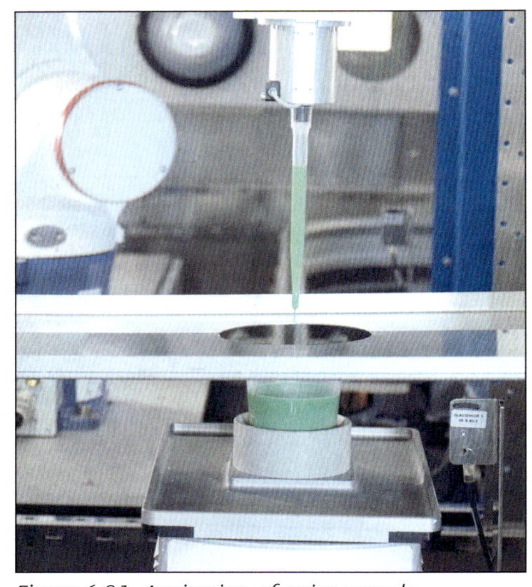

Figure 6.81: Aspiration of paint sample
Source: Chemspeed Technologies

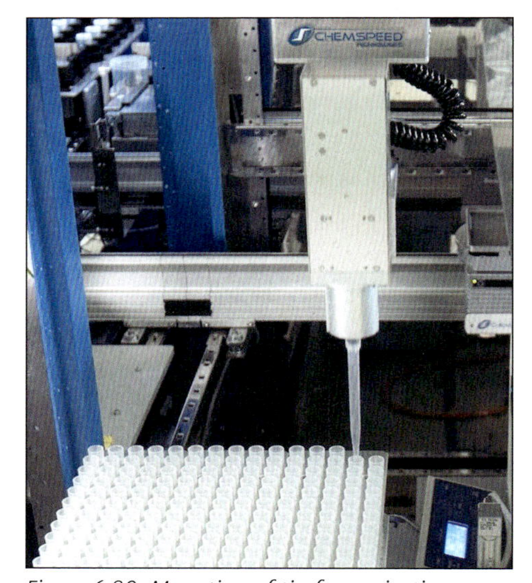

Figure 6.80: Mounting of tip for aspiration
Source: Chemspeed Technologies

Figure 6.82: Dispensing of paint sample during slow spinning
Source: Chemspeed Technologies

At the end of the dispensing process, the disposable tip is ejected and discarded over the waste container. The spin coater massively increases the rotation speed and thus produces a thin, very uniform coating film. The spin coating process is relatively short and requires very little coating material, but again depends on the size of the substrate. After completion of the coating process, the substrate is picked up again by the robot and sent to the next workflow step, classically storage in the substrate store for curing or heat activate curing in an oven, see Figure 6.83 and 6.84. Cleaning is not necessary during the spin coating process, which is why it is ideal for automation.

6.4 Curing methods

Curing of paints and coatings is the process of transforming the liquid or semi-solid film into a solid, durable layer through physical or chemical mechanisms. Curing enhances the coating's mechanical strength, chemical resistance, and adhesion to the substrate. There is almost a dozen of curing mechanisms [7, 8], but not all of them having importance in common laboratory automation. Thus, the focus in this chapter will be only on drying at room temperature, accelerated heat drying, heat curing and radiation curing.

6.4.1 Drying at room temperature

Although the term drying at room temperature triggers a minimal effort, the automation examples show various different realized approaches to make it efficient without demanding a lot of space. Automation with robots allows substrates to be stored in a very space-saving manner, which would not be possible with manual workflows. In addition, thanks to clever inventory management, the software always knows where in the warehouse which substrate with which coating film is located.

The Swiss company Chemspeed offers different storage systems and configurations for drying colour films at room temperature. The systems differ in the format of the substrate, but also in the height of the substrates. Aluminium labels and plastic films only have a very low height and can bend slightly under certain circumstances. This phenomenon also applies to all non-rigid substrates with a coating on one side. Rigid substrates such as glass or wood panels, on the other hand, can have a height of

Figure 6.83: Spin-coated paint sample
Source: Chemspeed Technologies

Figure 6.84: Pick up of coated substrate
Source: Chemspeed Technologies

a few mm. In addition to the height of the substrates, a free space in storage cassettes must also be taken into account, which is required for lifting and removing uncoated substrates and in reverse order when putting them back into storage. This can be seen in the Figures 6.85 and 6.86.

So-called shuffles are often used to retrieve and store the substrates, see Figure 6.87. The shuffles themselves move vertically or horizontally on a lift, depending on the configuration, in order to be able to move to all positions in the cassette. The cassettes can sit statically in a warehouse, but there is also the option of arranging them on a carousel. So that the substrates can be operated, the carousel with the desired cassette moves to the lift position with the shuffle. The shuffles can handle the substrates in the cassettes behind the transport lock. The substrates with the coated film on them will stay there until dried. Of course, the user can take the cassette out at an earlier stage if desired. The transport lock is required to prevent the substrates from accidentally slipping out when the operator removes a cassette from the automation system, see Figure 6.88. The

Figure 6.85: Pick up of a glass substrate from a three-module storage configuration with a total capacity of 900 substrates

Source: Chemspeed Technologies

Figure 6.86: Substrate positions in carousels are served with elevator shuffles

Source: Chemspeed Technologies

Figure 6.87: Single substrate is picked up for storage while drying

Source: Chemspeed Technologies

Figure 6.88: Dried drawdown on substrate removed by the operator

Source: Chemspeed Technologies

storage systems are comparable to a high-bay warehouse, just miniaturized.

Storage systems, regardless of the manufacturer, are not really rock science and are quite similar. At Füll Lab, substrate logistics are handled by a multi-axis robot. Instead of a shuffle, there are two long fingers at the end of the robot arm which move the substrate into the storage position of a cassette or a static storage system, see Figure 6.89. Here too, the substrate is inserted in a raised position during storage and then lowered into its final position by just a few mm. Füll Lab offers different configurations depending on the size and throughput of the HTE. A large storage system, for example, is operated with a multi-axis robot, which also moves on a linear axis to access all storage positions [9], see Figure 6.90. In common with other automation suppliers, the coated substrates stay in the storage until dried and then be taken out by the user.

Labman also has different storage systems in its portfolio – from small to large, depending

Figure 6.89: Substrate is stored for drying at room temperature Source: Füll Lab Automation

Figure 6.90: Large storage system for drying served by a multi axis robot on linear axis Source: BASF

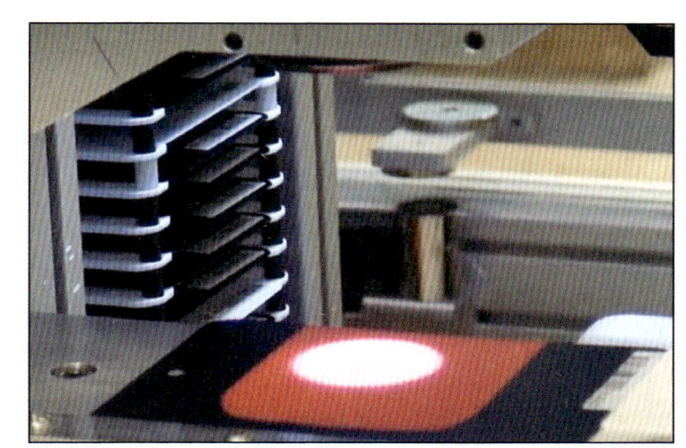

Figure 6.91: Substrate with wet film before transfer to storage Source: Labman Automation

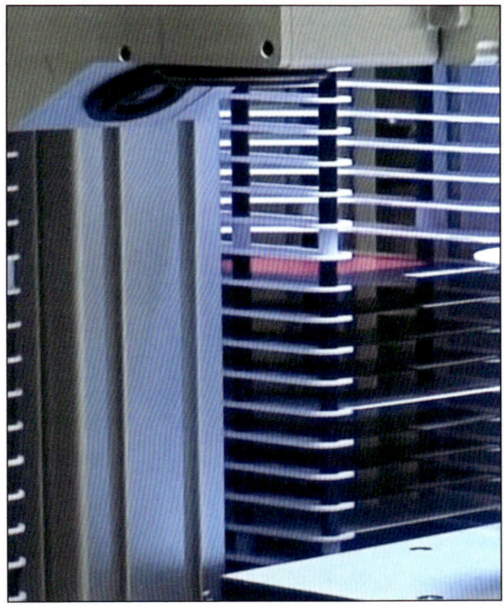

Figure 6.92: Tower storage system for drying of coating films Source: Labman Automation

Figure 6.93: Large storage system with cassettes for drying Source: Labman Automation

on the size of the HTE. Interestingly, Labman equips its systems with a static substrate storage system with different substrates, some of which are removed individually with a vacuum gripper and brought to the application, see Figure 6.91and 6.92. After application, these are pushed with a slider into a cassette that can be positioned at the right level along the z-axis. The cassette principle is therefore only used to store the applied films on the substrates, in which the drying takes place, see Figure 6.93. During or after drying, the applied substrates can be easily and quickly removed from the automated system.

6.4.2 Heat curing

Depending on the chemistry used in the paint, drying, or rather curing, of the paint can be accelerated or even initiated at a higher temperature. In contrast to manual loading of curing ovens, which can be handled relatively quickly, the process is somewhat slower with automation. In some cases, only individual substrates are loaded, but depending on the workflow, several substrates can be loaded in immediate succession. This causes longer opening times for the oven, which leads to a significant cooling of the temperature inside.

This is why a different furnace setup is required for automation. At Chemspeed, curing ovens in the industry are converted so that the

Figure 6.94: "Flex Heat Curing" with two ovens and transfer station Source: Chemspeed Technologies

Figure 6.95: Opening of oven drawer by multi axis robot while holding draw down
Source: Chemspeed Technologies

Figure 6.96: Positioning of fresh coated substrate in one of the oven positions
Source: Chemspeed Technologies

front door is replaced by a drawer system. As a result, only a small part of the oven is open during loading and unloading. Thus, the high temperatures, for example 180 °C, remain relatively stable and recover quickly after the drawers are closed. The highest temperatures that can be reached are 240 °C. As standard, 6 drawers are installed per oven and, depending on the size of the substrate, 2 to 4 positions are realized. This gives in total between 12 and 24 positions in the heat curing system per oven. If this is insufficient, the ovens can be multiple times integrated into a "Flexshuttle" , see Figure 6.94 and 6.95.

To transfer the fresh coated substrates into the oven, the tool head on the multi axis robot has a pin to open and close the drawers of the oven, see Figure 6.96. In addition, the tool head of the robot also has gripper fingers, which can grab the fresh coated substrates and put them into their positions for the curing process. If required, the system also offers a flash-off area before the substrates get loaded into the oven, see Figure 6.97. The reverse is also possible, means unloading the substrates after heat curing into the intermediate positions for cooling down. At ambient temperature, the substrates will be transferred to the storage system, which is also used for the substrate dried at room temperature.

The principle used for the thermal curing of wet paint films is comparable at Füll Lab. The implementation is slightly different. Füll Lab also tries to keep the opening time of the curing oven short so that the temperature inside does not drop too much. The oven does not have a classic door hinged on the side, but a vertical sliding door on the front of the oven, see Figure 6.98 and 6.99. This opens upwards and the 10 individual substrate positions become accessible for the multi-axis robot with the substrate. Of course, Füll Lab also offers intermediate storage for flash-off of the coating films before curing. When the curing period has finished, the sample are taken out and positioned in a rack respectively cassette to cool down and stored before taken out by the operator after the run is finished.

Figure 6.98: Transfer of coated substrate into oven Source: Füll Lab Automation

6.4.3 Accelerated drying on heating plates

Between drying at room temperature and baking the paints at high temperatures, there is the case of accelerated drying. This is often used

Figure 6.97: Intermediate storage for cooling down the substrates Source: Chemspeed Technologies

Figure 6.99: Sliding door of oven is closing Source: Füll Lab Automation

between 40 and 60 °C. Drying too quickly or at too high a temperature would have an impact on the optical properties of the lacquer. For example, the paint film at the interface with the air would close too quickly, resulting in poor flow or pin holes caused by too fast degassing or yellowing, to name just a few defects.

For this reason, Chemspeed has developed so-called heating plates. Contrast cards with draw-downed coatings are dried at 50 °C on the heating plates, see Figure 6.100 and 6.101. The only slightly higher temperature accelerates the drying process without any negative consequences. During the drying process, however, additional hold-down devices must be placed on the contrast cards by the robot. This ensures that the contrast cards lie flat and thus ensure good heat transfer and, on the other hand, prevents the card from curling due to the temperature and the fact that it is only coated on one side. Similar to the heat curing in ovens, the substrates will be transferred by shuttles to the storage area to finalize the run. The user is then able to unload the cassettes and collecting the drawdowns for his records.

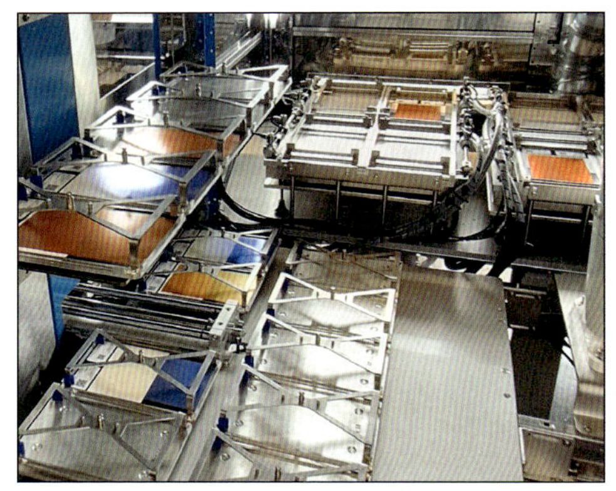

Figure 6.100: Accelerated drying station with heat plates and hold-down devices

Source: Chemspeed Technologies

Figure 6.101: Heat plate station for 4 substrates

Source: Chemspeed Technologies

Figure 6.102: UV curing module as part of the "Flexshuttle" concept

Source: Chemspeed Technologies

Figure 6.103: Stationary UV dryer is available with different lamp types

Source: Chemspeed Technologies

6.4.4 Radiation curing

A not so common method for drying in automated laboratories is UV curing, one of the disciplines of radiation curing. There, ultraviolet (UV) light initiates a polymerization reaction in a UV-sensitive resin, for example used in UV-curable inks, wood finishes, and electronic coatings. Chemspeed and Füll Lab have developed solutions for an automatic horizontal UV curing and drying system for flat substrates. The Chemspeed UV solution includes 2 lamps: 1 standard Hg lamp and 1 Hg Ga-doped lamp. The spectrum of available UVH lamps reaches from standard (mercury) over type Ga (gallium doped) to type F (iron doped). Two different setups are available for substrate formats of 85 x 150 mm or 204 x 284 mm, both maximum 10 mm thick, see Figure 6.102 and 6.103.

Optional items are the module enclosure for contained atmosphere with two shuttle air locks for shuttle arrival and shuttle departure, closed ceiling and tray for "FLEX" module. In addition, the module can be upgraded to handle LED lamps. Immediately after curing, the substrates will be processed to a possible first testing and characterization workflow step or transferred to the earlier described storage systems.

6.5 Literature

[1] CHALLENER, C.; "Industrial Paint Application Technology: An Overview", p. 50ff, JCT Coatings Tech Magazine, April 2004

[2] BROCK, T.; GROTEKLAES, M.; MISCHKE, P.; "European Coatings Handbook", p. 285ff, Vincentz Network, Hannover, 2010

[3] TROSTLE, J.; "Best Application Technology", p. 68ff, JCT Coatings Tech Magazine, April 2012

[4] GOLDSCHMIDT, A.; STREITBERGER, H.-J.; "BASF Handbook on Basics of Coating Technology", p. 477ff, Vincentz Network, Hannover, 2018

[5] Füll Lab Automation, "From Bosch Lab Systems to Füll Lab Automation", www.fuell-labautomation.com/company/history, 2024

[6] HECKER, M.; "Kleinste Tropfen im Visier", p. 56ff, CITplus, Wiley-VCH, Weinheim, 5-6/2020

[7] BROCK, T.; GROTEKLAES, M.; MISCHKE, P.; "European Coatings Handbook", p. 318ff, Vincentz Network, Hannover, 2010

[8] GOLDSCHMIDT, A.; STREITBERGER, H.-J.; "BASF Handbook on Basics of Coating Technology", p. 584ff, Vincentz Network, Hannover, 2018

[9] BASF, "Resins for Coatings Applications: Speeding up innovation with digital material profiling", https://youtu.be/S6CIuP4wv_U?si=uMJBiYKBtRKdVMLw, 2021

7 Technologies for testing and characterization

Paints and coatings are tested for their properties to ensure that they meet the desired requirements and standards. Property testing is crucial to ensure the functionality, durability and safety of products. A complete listing of methods for testing wet paint properties and applied paint films is beyond the scope of this book. At this point, reference should be made to specialist books that have also been published by Vincentz Network [1–3]. For this reason, the Table 7.1 lists frequently recurring tests and inspections.

Paint tests are often conducted according to international standards for consistency and reliability such as ISO (International Organization for Standardization), ASTM (American Society for Testing and Materials) International, National standards, e.g. DIN (German standards), BS (British standards), SN (Swiss standards), etc. The list of tests and checks has grown historically. This means that there are still many older methods, some of which are not exactly automation-friendly. In particular, these are purely mechanical tests that do not provide digital results. Advantageously, many manufacturers now offer revised methods in a digital version, which can be integrated more elegantly into automation systems and connected via interfaces.

As the list of tests is significantly larger than the elements for dosing raw materials, manufacturing paints and coatings and their application, the choice for investing in automation for test methods is very budget relevant. The investment criteria for this essentially consist of the largest bottleneck for tests and the frequency of their use for the intended automation workflows. For example, it makes economic sense to invest in characterization methods that are almost always (>80 %) carried out in all workflows. Tests that are only carried out for a maximum of every 5th sample (<20 %) are not economically attractive and should continue to be carried out manually.

Table 7.1: Examples of frequent paint and coating characterizations

Coating material	Wet coating film	Dry coating film
pH	Drying time	Thickness
Viscosity/rheology	Film formation/levelling	Colour/imaging/dry opacity
Pot life	Sagging	Gloss/haze/reflectivity
Volatility	Wet opacity	Levelling
Solids content	Open time/workability	Hardness (pendulum, pencil, nano indentation
Density	Rub out (floating)	Flexibility (mandrel bending, cupping, elongation
Dispersibility/particle size	Defoaming	Resistance (water, chemical, corrosion, scrub, stain, fire, scratch, abrasion)
Anti-microbial	Shrinkage	Accelerated weathering (degradation, chalking, micro bacteria)
Degassing	Mud cracking	Adhesion (crosscut, shear, stone chip)
Stability (syneresis, sedimentation)/turbidity	Surface tension	Surface tension (contact angle, water repellancy)

However, the required accuracy and reproducibility can be an exception. Otherwise, it can also be dismissed as a "nice to have". To summarize, this is the point that causes the most difficulties in the investment decision, what's in and what's out. But in the end, as is so often the case, the available budget is the deciding factor.

7.1 Characterization of wet coating materials

Wet paints undergo the first tests during their production, before they are finished. These include, for example, fineness after dispersion, measurement of particle size, pH and possible correction. Further tests can be carried out immediately after completion of the formulation, for example flow properties. Other tests are only carried out after weeks of storage to check stability.

7.1.1 Fineness of grind

The automated determination of dispersibility by a Hegman gauge was developed over 25 years ago by Labman. Since then, the "Tidas" has gone through several iterations, each honing its performance and reliability. The measurement can be carried out semi-automatically as a standalone, i.e. the operator only has to drop the paint sample onto the gauge, insert it into the measuring device and mount the blade (Figure 7.1 to Figure 7.2).

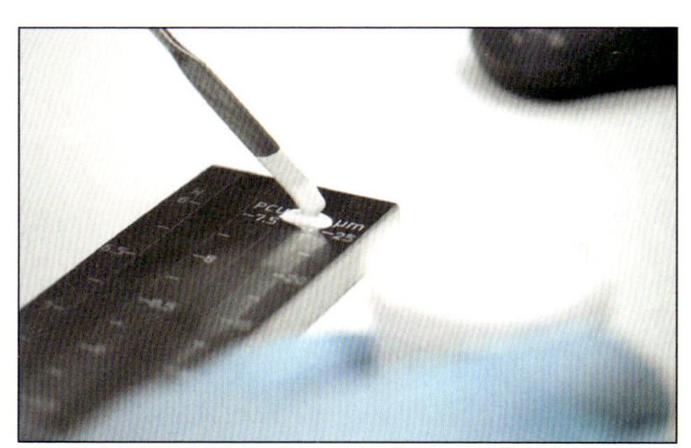

Figure 7.1: Manual dispensing of paint onto Hegman gauge Source: Labman Automation

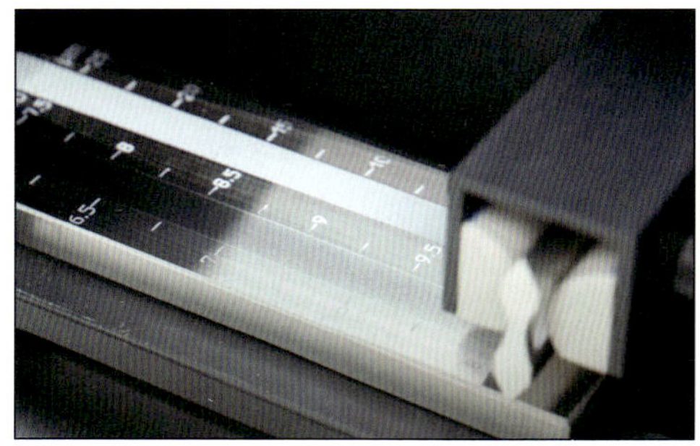

Figure 7.2: Automated application of paint by blade Source: Labman Automation

Figure 7.3: Digital reading of Hegman fineness Source: Labman Automation

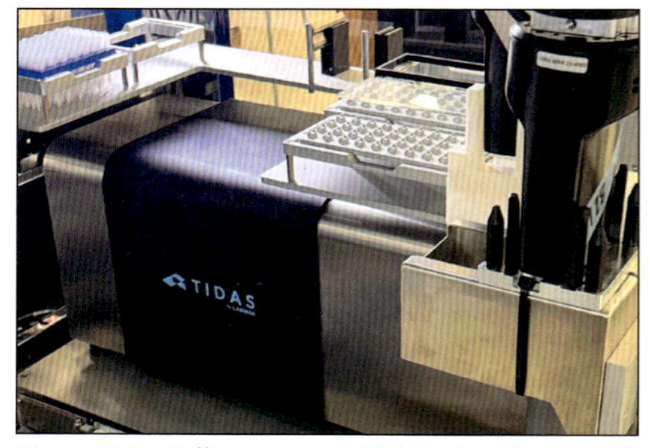

Figure 7.4: Fully automated Tidas by Chemspeed Source: Chemspeed Technologies

The rest of the test is done automatically plus the inspection by the digital camera and the evaluation of results. The system integrates a high-definition line scan camera, which contributes to the generation of distortion-free images. After the Hegman gauge is automatically drawn down, the system shines a beam of light onto the dispersion to allow the image to be captured (Figure 7.3).

The software is designed to eliminate dust, bubbles and foaming from the dispersion. It can also robustly take readings from clear and metallic materials which are notoriously difficult to rate by eye, owing to the dark field illumination technique used to capture the image. The dark field illumination technique is used to capture an image of the Hegman gauge surface. This replicates the method used for manual inspection, as outlined in ASTM D 1510 and ISO 1524. With this technique, a beam of light is directed at a low angle onto the gauge surface. This reflects at the same angle off the flat surface when no particles are present. Once particles are visible, the light scatters and is detected by the camera. This creates a high-contrast image. The software package empowers users with a versatile toolkit for tailored operation and analysis. The raw image captured is presented alongside the digital analysis, and the micron count. The count identifies the exact number of particles per micron slice, which provides a defined and repeatable rating. Users are able to setup Tidas to automatically produce a pass or fail against their quality standards. Finally, only the gauge and the blade need to be cleaned by the operator for their next use.

A fully automated solution for measuring fineness of grind is available from Chemspeed (Figure 7.4). They have gone one step further and integrated Labman's "Tidas" into the "Flexshuttle" concept so that manual steps are no longer necessary. This means that a sample is aspirated directly from the formulation vessel using the multi-dispenser with disposable positive displacement tips and dispense onto the gauge. The gauge and the blade are operated by a robot and are also cleaned and dried automatically after application. The offered setup can measure more than 250 samples without interruption due to the generously designed storage for the positive displacement tips. Both aqueous and solvent-based paints and coatings can be tested. Depending on the configuration, an active washing station for aqueous and/or solvent-based systems is integrated.

7.1.2 Particle size

Dynamic light scattering (DLS) is a widely used technique for measuring particle size in suspensions, emulsions, and solutions, particularly for particles in the nanometer to micrometer range. Typical applications of DLS are the characterization of nanoparticles and colloidal particles, particle/droplet determination of polymer solutions, emulsions and dispersions as well as for quality control of paints, inks and coatings. The offer of particle size instrument of different manufacturers is broad, e.g. from Malvern, Horiba, Anton Paar and Microtrac Retsch, etc. One example of the last-mentioned manufacturer is the "Nanotrac" Wave II, a highly flexible DLS analyzer which provides information on particle size, zeta potential, concentration, and molecular weight (Figure 7.5 to Figure 7.6). The measuring range of particle size is from 10 nm to 10 µm and has an enhanced zeta potential capability from -200 to +200 mV.

The aspiration of the sample is performed by an overhead XYZ-axis robot and liquid dispensing setup with disposable positive displacement tips. The measurement requires a minimized sample volume, which is dispensed into the integrated measurement cell to allow automated dilution with either water or solvent. The system is able to measure in a wide concentration range from ppb to 40 %. During the measurement the disposable tip is dropped into the waste container. The automated cleaning of the cell is performed after the completed measurement. Examples of automated particle size measurement systems can be found by Füll Lab and Chemspeed.

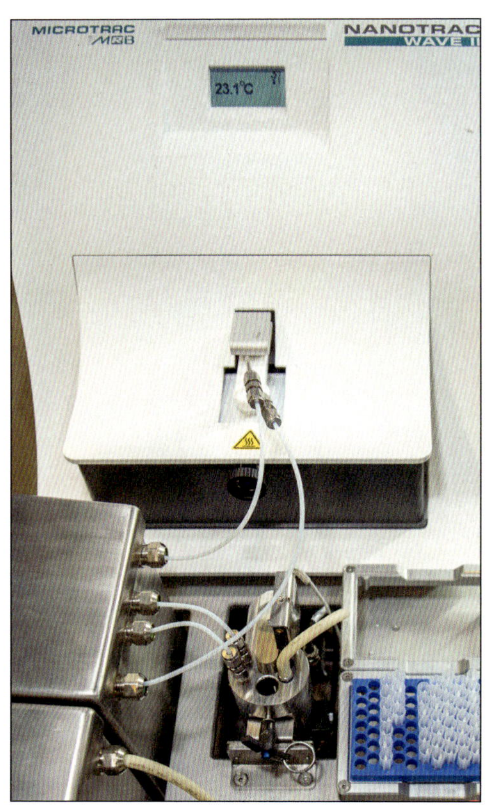

Figure 7.5: Automated particle size setup with the "Nanotrac" Wave II Source: Chemspeed Technologies

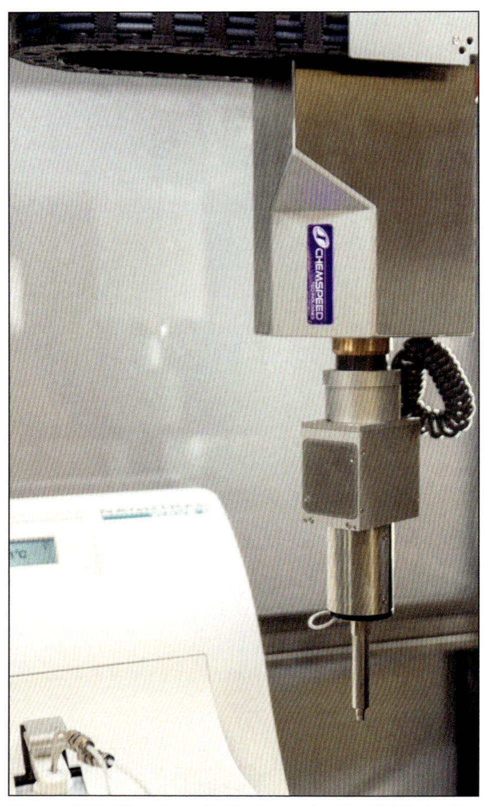

Figure 7.6: Overhead robot nearby the PSD device Source: Chemspeed Technologies

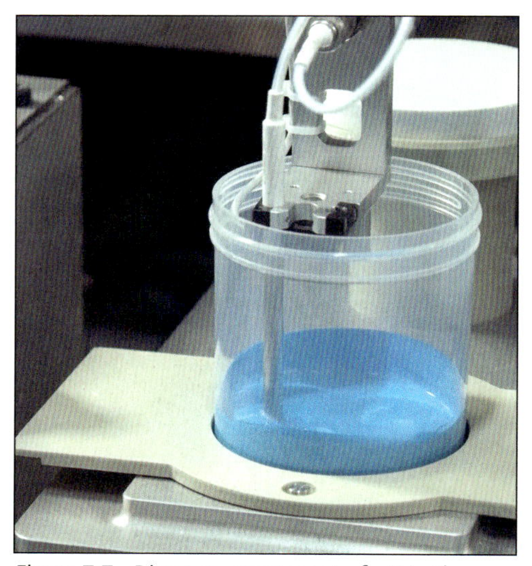

Figure 7.7: Direct measurement of pH in the sample vessel Source: Chemspeed Technologies

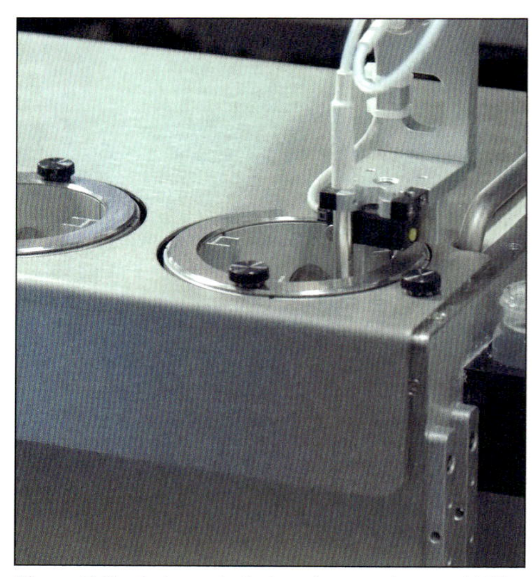

Figure 7.8: Automated cleaning station and KCl storage for the pH probe Source: Chemspeed Technologies

7.1.3 pH measurement and adjustment

The next test is very likely to be the pH during production, provided it is an aqueous colour system. This measurement is mandatory, which is why it is offered by all automation solution manufacturers in their respective concepts. However, the implementation or integration always looks different. Depending on the manufacturer, the electrode is integrated into a setup for dispersing or separately as a physically separate workflow step. This has the advantage of reducing the cycle times in one module because it is carried out in two separate modules in parallel.

The described pH measurement example of Chemspeed is carried out on-line directly in the formulation vessel. When not in use, the probe is stored in a KCl solution, as for manual use. The measuring range extends from pH 0 to 14 with an accuracy of ± 0.1. Depending on the formulation, the filling level in the container can vary, which is why a liquid level detection is also integrated. An active washing station is used to clean the pH probe, which is located in the immediate vicinity. It is also possible to integrate an overhead stirrer for ensuring solution homogeneity and accurate measurement (Figure 7.7 to Figure 7.8).

Complementary to pure pH measurement, manufacturers also offer pH adjustment with various reagents as an automated solution. In addition to the pH probe, a stirrer for homogenization and a dosing needle for adding the reagent solutions, for example sodium hydroxide or amino alcohols, as the coating materials are usually anionic aqueous systems, are used at the same time (Figure 7.9 to Figure 7.10).

7.1.4 Viscometry

Immediately after the completion of paints and coatings, the flow behaviour of the wet sample is measured in the vast majority of cases. This can be a simple determination of the viscosity by means of rotational measurement at one or more shear rates. For more in-depth statements on flow behaviour, a rheology measurement can be carried out over a wide shear band in rotation or oscillation. The portfolio knows hardly any limits. The fully automated systems integrate the viscometers and rheometers available on the market from well-known companies such as Ametek

Figure 7.9: In-situ pH adjustment while stirring
Source: Chemspeed Technologies

Figure 7.10: Automated cleaning station for pH probe and stirrer
Source: Chemspeed Technologies

(Brookfield), Anton Paar, TA Instruments, Thermo Fisher and others. The fully automated systems also have solutions for standalone solutions, as does Anton Paar as a manufacturer of viscometers and rheometers.

Standalone solutions for measuring flow properties are becoming increasingly popular. This is not particularly surprising, as manual measurement can be classified as dull routine work without any particular intellectual excellence but is nevertheless very necessary. Standalone solutions for measuring viscosity have one to several viscometers within the platform, depending on the required throughput. Instruments from Ametek and Anton Paar are widely used, analogous to manual laboratory operation. For automation operation, they are mounted on at least a simple z-axis, but sometimes also on XYZ-axis robots.

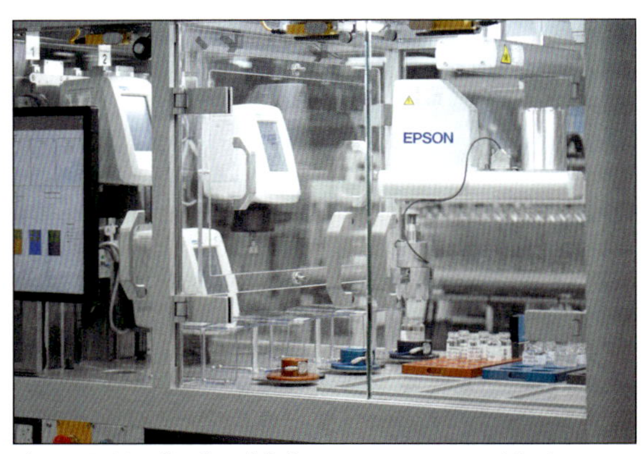

Figure 7.11: "Helipath" viscometry system with three Brookfield viscometer Source: Labman Automation

Figure 7.12: Viscometry setup with "Rheolab QC" Source: Chemspeed Technologies

A tidy solution with three Brookfield devices is offered by Labman (Figure 7.11). The system contains three viscometers that take measurements of three different size sample jars. The jars are loaded into the system and their locations are entered on the UI. Once the user presses start the robot arm picks the sample, scans the barcode, and places it into one of the appropriate shuttles. The shuttle then moves under the viscometer where a measurement is taken. Once the measurement is finished the shuttle moves back to the robot arm and the sample placed back to the rack. The viscometer lowers into a wash bath where the geometry is washed and dried. The viscometers can be setup to use either all the same jar type or a combination of different types. Shuttles on magnets and quick release airlines and racks are colour coded as to provide clear feedback on which jar size each viscometer is setup to use.

The Swiss manufacturer Chemspeed offers different variants for measuring viscosity. In addition to a standalone solution, a functional module can also be integrated into a "Flexshuttle" system. The rotational viscometer, for example "Rheolab QC" from Anton Paar, is rigidly mounted on the overhead robot and moves on three axes within the system (Figure 7.12). As soon as the sample reaches the module via shuttle, the measuring geometry is immersed in the sample. The measurement with the fully integrated "Rheolab QC" from Anton Paar covers the low to medium viscosity shear range. The system is designed so that the viscosity can be set in addition to the viscosity measurement. Two different liquids can be selected from a reservoir for dosing in order to set the desired viscosity. After measurement and, if necessary, viscosity adjustment, the measuring geometry is actively cleaned in the integrated washing station, which is either for aqueous and/or solvent-based systems available. When using solvent-based systems, an F90 safety cabinet is also integrated for storing the solvents.

For the measurement and adjustment of viscosity the modules of Füll Lab allow direct dispensing of liquids such as acids or bases from "BLS Syringes" and mixing of the formulation with an overhead stirrer. The automation of viscometry has been successful implemented. The sample in the formulation cartridge will be transferred beneath the viscometer, which will then move on its z-axis into the cartridge. After measurement respectively adjustment of the viscosity the viscometer moves up followed by removing the sample cartridges. Finally, the viscometer geometry will be automatically cleaned.

7.1.5 Rheology

The integration of benchtop rheometer models from TA Instruments, e.g. the HR-10 "Discovery Hybrid Rheometer", or "MCR 302e" from Anton Paar is standard at Chemspeed, but alternatively there is also the overhead version "DSR 502" from Anton Paar available. It is certainly worth mentioning that the use of a stand-alone granite antivibration setup for rheometer for stable operating conditions and high precision measurements is standard. This prevents the transmission of possible vibrations from other moving components, especially the robot, to the rheometer during measurement operation. The features of the automated rheometers do not differ from their

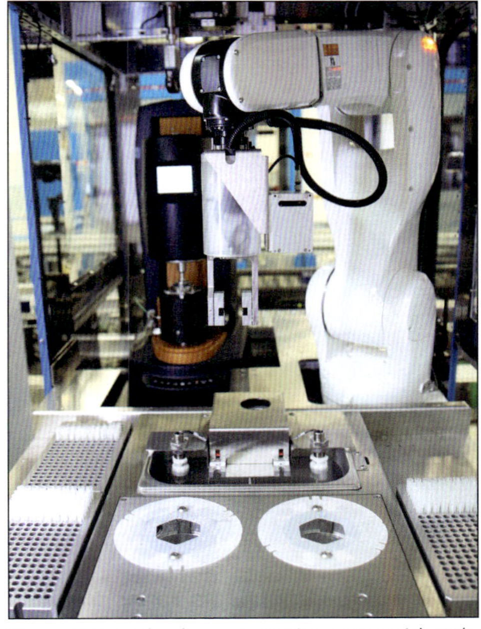

Figure 7.13: Rheolgy automation setup with TA's "Discovery HR-20" Source: Chemspeed Technologies

Figure 7.14: Automated sample dispensing on lower plate geometry Source: Chemspeed Technologies

manual operation. All three measuring systems are available in different dimensions of plate/plate, plate-cone and bob-cup in conjunction with a vacuum Peltier plate (air cooled circulator and temperature control system for 0 to 200 °C), solvent trap and trimming tool. For plate/plate and plate/cone, the system is supplied with two upper and two lower measuring geometries as standard, so that the parallel measuring operation and cleaning of the previous geometries can be done. When using bob/cup, only 2 bob geometries are required, but the number of cups must be matched to the number of samples (up to a maximum of 150 per run). Screw capper and pH measurement are offered as options.

The "Flex Rheo" module (Figure 7.13) is essentially operated by a multi-axis robot plus a vertical linear axis when using the screw capper. The measuring geometries are mounted on the rheometer using the multi-axis robot and the sample vessel, identified by a barcode reader, is then opened using the screw capper. Depending on the dimensions of the sample vessel, the number present on the system can reach above 100. The robot then retrieves the disposable tip from the storage with a maximum capacity of 273 tips, aspirates the sample and dispenses it precisely onto the measuring geometry (Figure 7.14). When using cups, more material is required for measurement, which is why the system is then equipped with up to 250 syringes of 10 ml. After the measurement, the measuring geometries are removed from the robot one after the other and parked in the cleaning station. Then the next experiment begins and during the measurement, the geometries from the previous experiment are cleaned and dried in parallel. When measuring samples with very high viscosity, a so-called paste dispenser with a tip cutter function, for widening the nozzle size for easier flow, is used.

A highly comparable setup for automated measurement comes from rheology specialist Anton Paar. "HTR 7000" (Figure 7.15) can be equipped with up to 96 measuring geometry units and also offers three different rheological geometries. The standard rack configuration offers space for 96 samples. Different types of racks for pipettes, syringes and trimming blades as well as temperature-controlled sample storage (up to 4 °C) are available. Thanks to mechanical decoupling of the rheometer and the automation unit, HTR achieves laboratory quality and is designed

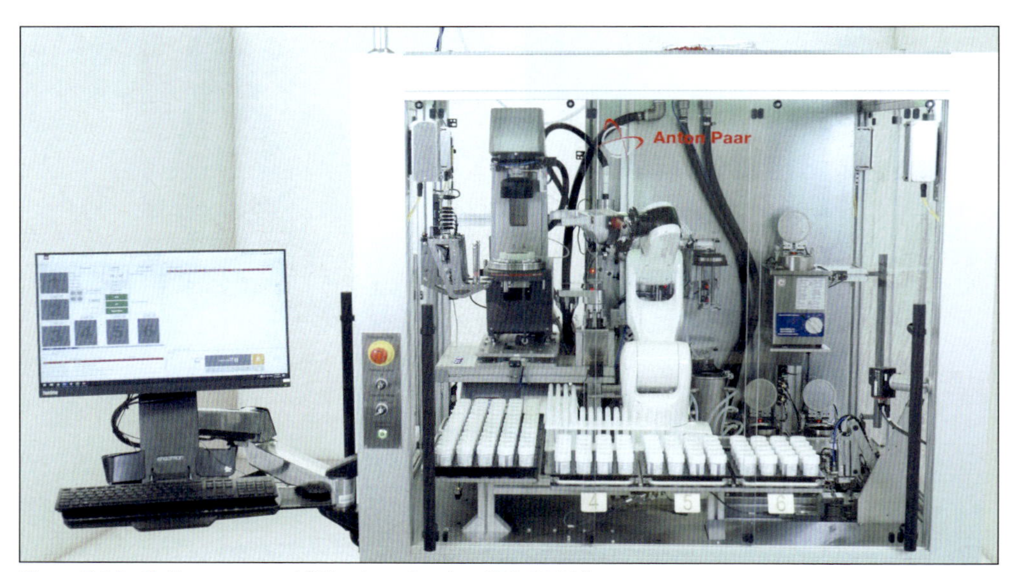

Figure 7.15: Fully automated "Rheology Station HTR 7000" *Source: Anton Paar*

for maximum efficiency. The "HTR 7000" comes with the high-end device "MCR 702" "Multidrive" and also offers a large number of features such as pH measurement, sample preparation, automated active multi-chamber cleaning, software connection, priority samples, integrated trimming tool with visual verification of trimming results. For the sample transfer, the sample vessels are opened using a screw capper and closed again later. The sample system is precise and can dispense volumetrically and gravimetrically (Figure 7.16). Syringes are used under normal circumstances. Even highly viscous systems can be transferred by automatically expanding the pipette openings.

Standard modules are used at Füll Lab for measuring liquid or pasty formulation. Typically, these modules integrate commercially available metrology devices such as a rheometer. A nice example is the integration of the overhead rheometer "DSR 301" (Figure 7.17) in the Integrated Lab Station or the newer version "DSR 502" as standalone "RheoModule" (Figure 7.18). The concept is quite similar to their viscometry automation. The sample cartridge is positioned beneath the rheometer, which when moves the geometry downwards in the sample. In addition, liquids can be added for the adjustment of the flow behaviour to avoid application problems. Again, after the completed measurement the geometry of the rheometer will be automated cleaned while the sample is transferred to the next workflow step.

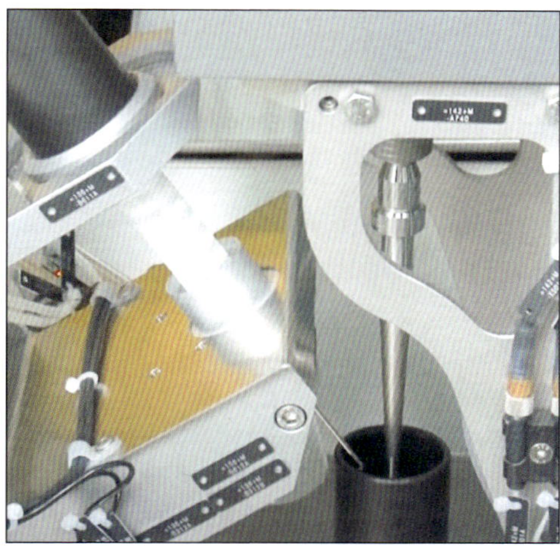

Figure 7.17: Overhead rheometer setup with "DSR 301" *Source: Füll Lab Automation*

Figure 7.16: Automated dispensing on lower geometry of "MCR 702" *Source: Anton Paar*

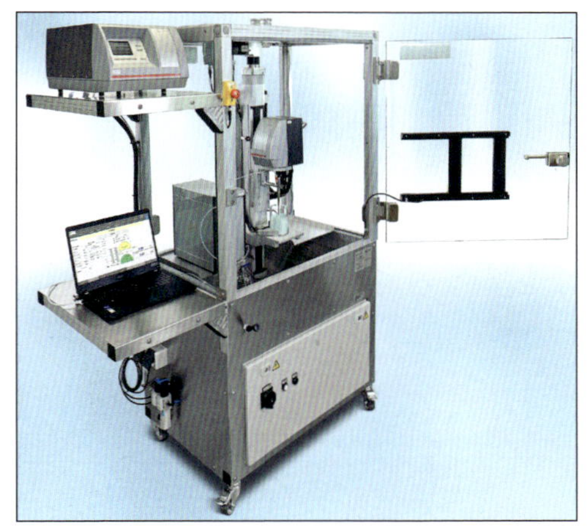

Figure 7.18: Standalone module with overhead rheometer "DSR 502" *Source: Füll Lab Automation*

7.1.6 Density

The determination of classic paint and coating parameters are of course also available as automated tests. The first example of this is the measurement of the density of wet paints. Paint pycnometers or a density ball are often still used for manual gravimetric determination. These methods are less suitable for automation. However, on the market are different solutions for the automated measurement of density available, e.g. by Chemspeed Technologies and Füll Lab Automation. With the integration of high-precision and very fast measuring analytical density meters, such as the Anton Paar "DMA 4500 M", a sample volume of just 1 ml is sufficient for measurement (Figure 7.19). The covered measuring range of 0 to 3 g/cm^3 is completely adequate and the precision of the digital resolution is 0.00001 g/cm^3. Using oscillating U-tube technology based on the patented pulsed excitation method, measurement results are obtained after 30 s and comply with ASTM D4052, D5002, D1250, ISO 12185 standards. The automated process starts after the wet paint reaches the module and is transferred to the parking station. The sample head is guided by the robot to the sample vessel. There the sample is aspirated and transferred to the density meter. Once the measurement process has started, the sample head is transferred to the cleaning station. Depending on the paint system being measured, cleaning is performed separately in the aqueous or solvent-based cleaning station, which is equipped with a sufficiently large storage container.

7.1.7 Solids content

There are various methods used in laboratory automation to determine the solids content. For the automation common moisture analyzers, e.g. from Sartorius or Mettler Toledo, are qualified for integration (Figure 7.20 to Figure 7.21). Classically, aluminium pans are transferred from the warehouse with space for 182 pans to an analytical on-deck balance with a resolution of 0.1 mg using the multi-gripper of the multi axis robot. Next, the robot picks up a 1 ml disposable positive displacement tip with its double tool and aspirates the wet paint from the sample vessel. This is then weighed into the aluminium pan. The glass doors are closed to prevent fluctuations due to draughts in order to precisely determine the weight. The aluminium pan is then transferred to the oven with a drawer system. The use of drawers prevents greater heat loss during loading compared to the use of a large door. Typically, the solids content is determined at 105 °C. However, the furnace used is also designed for temperatures of up to 240 °C. After the defined storage time, the aluminium pan is removed from the furnace drawer and placed in an intermediate storage area to cool down. After cooling, the aluminium pan with the residue is weighed again to determine the solids content of the coating. After weighing, the aluminium pan and the residue are discarded.

Figure 7.19: Automated density measurement setup
Source: Chemspeed Technologies

7.1.8 Tint strength

A large number of colour shades are available on the market for decorative architectural paints, but also for OEM and industrial coatings. It must therefore also be possible to determine the tinting strength automatically. For this purpose, colour pastes, sometimes in very small quantities, must be dispensed into the base colour with high precision. Füll Lab uses its volumetric dispensing system for liquids for this purpose, for example as part of the "Integrated Lab Station" or "Compact Lab Station". The formulation cartridge is transferred to the dispensing station and the colour paste is dispensed volumetrically (Figure 7.22). The sample is then homogenized, usually with a DAC, before application.

A quite similar approach is applied with the automation solution of Chemspeed. The reference sample of a base colour is compared with the manufactured base sample to be tested. This process can be carried out easily and precisely by using two dosing units with 240 ml cartridges for the colour pastes. The fully integrated multi-tool on the robot arm can thus handle the

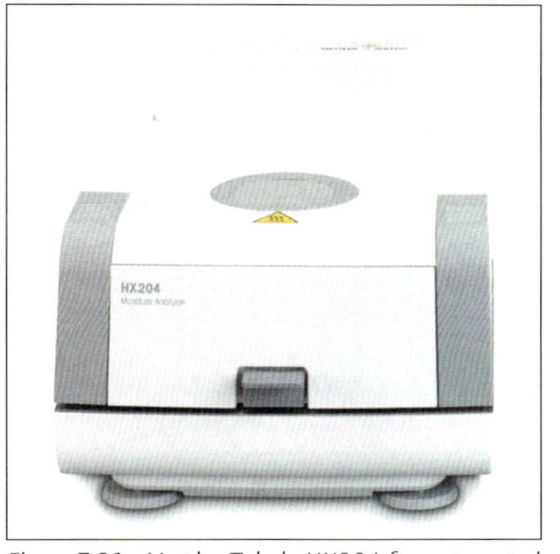

Figure 7.21: Mettler Toledo HX204 for automated solids content determination Source: Mettler Toledo

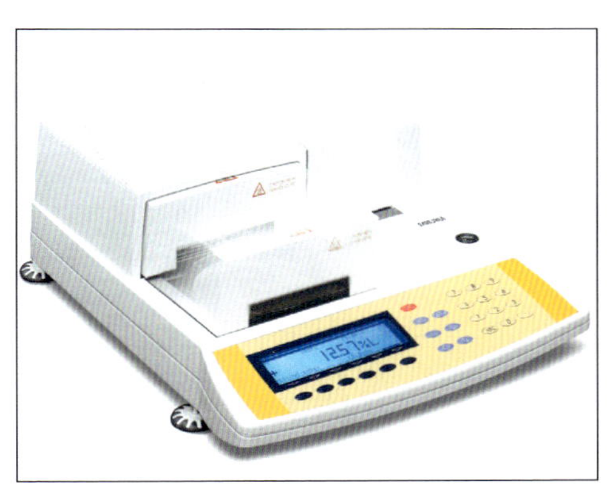

Figure 7.20: Sartorius MA100 for automated solids content measurement Source: Sartorius

Figure 7.22: Volumetric dispensing of colour pastes into sample container Source: Füll Lab Automation

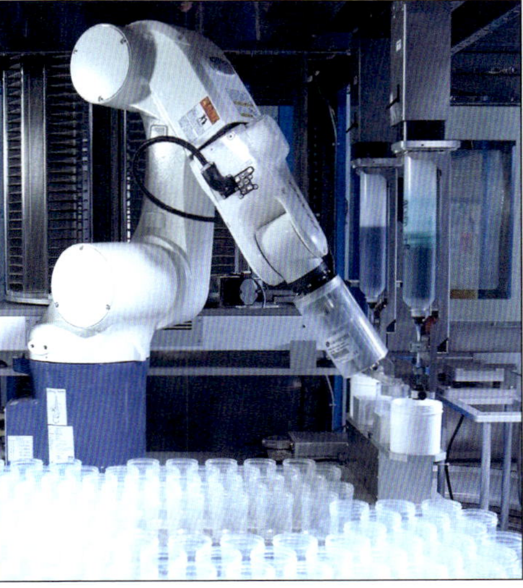

Figure 7.23: Automated tint strength module with balance, colour paste containers, vessel storage and underfloor DAC Source: Chemspeed Technologies

sample transport, vessels, substrates and, if necessary, with tool extension for further processes. The system uses 60 ml vials as standard from a storage system with a maximum of 124 vials (Figure 7.23). The colour pastes are dosed gravimetrically on an on-deck balance with 1 mg resolution. After weighing, the vials are loaded into a DAC for homogenization of the base colours and colour pastes. The mixing speed can be variably increased in steps or set at the same speed. At the end of the mixing time, the two vials are unloaded again and transported to the application, e.g. drawdown.

7.1.9 Storage stability/turbidity

Testing the storage stability of paints and coatings can include various elements. In addition to the change in flow properties, wet paints can segregate and form phases. These can manifest themselves as syneresis or sedimentation and sediment formation. Alternatively, the turbidity of clear solutions or clear coats can also be determined. Their characterization is carried out during storage over various periods of time. For qualitative and quantitative determination, multiple light scattering measurement technology is used in laboratory automation, for example with the "Turbiscan Lab" from Formulaction, which is integrated by Chemspeed and Füll Lab. The following example presents the automated module of Chemspeed using an overhead robot with arm extension on a linear axis system responsible for logistics within the system (Figure 7.24 to Figure 7.25). The robot uses the multi-gripper to transport the glass cells with modified polycarbonate screwed top cap and butyl/"Teflon" seals between the storage and the "Turbiscan". Another tool available is an analytical overhead gravimetric dispensing unit (GDU), which is used to aspirate the wet sample into the glass cells. As soon as the wet sample arrives, the workflow within the system starts. The GDU grabs a syringe from the storage and aspirates the wet sample from the sample vessel. It is then precisely weighed undiluted into one or more glass sample cells. After the tool change, the multi-gripper transfers the sample cell to the "Turbiscan" for measurement of sedimentation, flocculation and creaming stability. The sample cell is then transferred to an SBS format rack in the warehouse. Once the run is complete, the operator can manually remove the filled racks from the system and place them in the external storage stability ovens. After one or more defined storage periods, the racks with the sample cells are returned to the system by

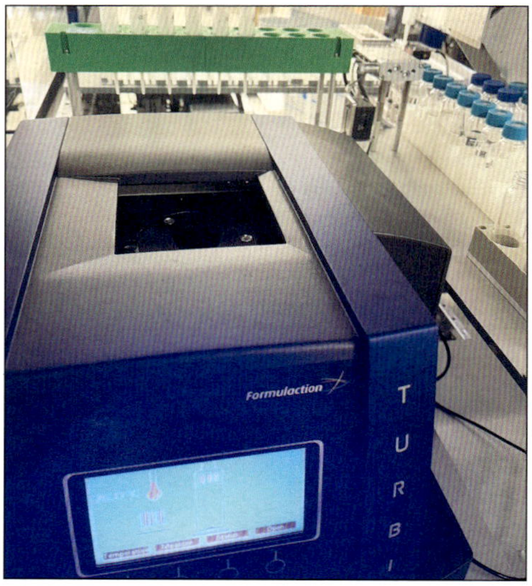

Figure 7.24: Automated "Turbiscan" in storage stability module Source: Chemspeed Technologies

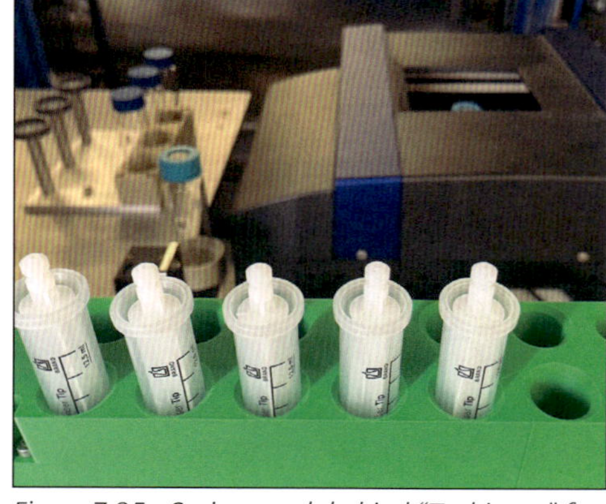

Figure 7.25: Syringe rack behind "Turbiscan" for sample aspiration Source: Chemspeed Technologies

the operator and the measurements are repeated in the "Turbiscan". This process is repeated until the defined end of the storage stability test. The qualitative and quantitative results are transferred by the software to the database for further evaluation.

7.1.10 Foaming

A rather special test for wet paints is the behaviour of foam formation and its stability or defoaming behaviour, especially with water-based paints and coatings [4]. Automation is here ideal for ensuring reproducibility and obtaining comparable results. A minimum quantity of 50 ml is required to carry out the measurement. Another basic requirement for characterization is a transparent sample vessel, for which glass vials are best suited. In the example described of Chemspeed, the sample arriving on the shuttle is fixed in the parking position. The foam in the sample vessel is generated by the high-speed stirrer moving in on a linear axis. The resulting foam and its decomposition (Figure 7.26 to Figure 7.28), as soon as the stirrer is stopped, is optically determined by means of LED illumination at a defined wavelength. The agitator is then cleaned in the integrated active washing station in an aqueous or solvent-based medium, depending on the paint system.

Figure 7.27: Paint sample before stirrer is stopped Source: Evonik

Figure 7.26: Paint sample before start of foam test Source: Evonik

Figure 7.28: Paint sample is starting to defoam after stop of stirrer Source: Evonik

7.1.11 Surface tension

Tensiometers are the choice for measuring the surface tension of liquid samples. The ring tear-off method is a variant and historical predecessor of the Du Noüy ring method as it is used today for measuring surface tension or interfacial tension. Instead of repeatedly detecting the maximum force by cyclic elongation and relaxation of the lamella as is usual with modern tensiometers, the lamella is overstretched until it breaks, so that only one value is determined per measurement. Füll Lab as well as Chemspeed do offer the integration of tensiometers, covering the surface tension range from 10 mN/m to 100 mN/m (dyn/cm), for the automated handling. The company Krüss, a specialist for determing the surface tension of liquids and solid surfaces, offers pure standalone tensiometer for the automated measurement with their new generation tensiometer "Tensiio" (Figure 7.29 to Figure 7.30). The number of applications is large, for example the surface tension and interfacial tension measurement for wetting and cleaning agents, emulsifiers, or other surfactants.

7.1.12 Integration of analytical devices and other methods

Analytical measurement methods can also be incorporated after the paint has been manufactured to characterise chemical properties. These serve various purposes to ensure the quality, performance and composition of paints and to evaluate their suitability for specific applications.

- Electrical conductivity to determine the ease of electrostatic application, as a measure of the condition and purity (stability!) of water-based coatings and for the stability of electrodeposition coatings.
- Near infrared spectroscopy (NIR)
- Gas chromatography (GC) with mass spectrometer
- Titrations for acid value or hydroxy value
- Differential scanning calorimetry (DSC)
- Gel permeation chromatography (GPC)
- Nuclear magnetic resonance (NMR)

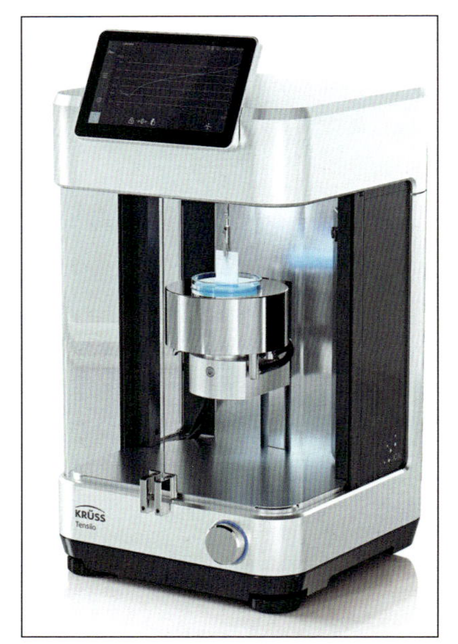

Figure 7.29: Automated standalone device for the measurement of surface tension of liquids Source: Krüss

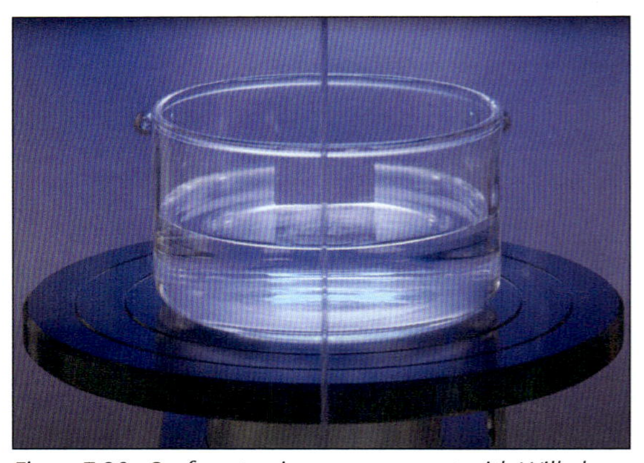

Figure 7.30: Surface tension measurement with Wilhelmy plate Source: Krüss

7.2 Characterization of wet coating film

Knowledge of the behaviour of freshly applied paints and coatings is important in order to qualify their suitability. This essentially concerns their processing properties and behaviour until film drying or film curing occurs. Some of the manually applied methods are only automated in their execution. The situation is different with specially developed methods that are intended to reproduce the technical application behaviour. Here, a certain correlation to practice must be demonstrated.

7.2.1 Sagging

Testing sagging is particularly important for paints and coatings that are applied manually to vertical surfaces. Good sagging resistance, even with high wet film thicknesses, provides the user with a high level of application reliability and prevents unsightly runs and/or curtains on the wall surface. For objective testing, a comb doctor blade is used in the same way as for manual testing. Identical to the application by draw down described above, after the automated dispensing of the paint onto the substrate, the comb doctor blade is moved over the substrate by the robot (Figure 7.31). After application, the comb doctor blade is taken to the washing station for cleaning. The substrate is then moved from the horizontal application table by the robot arm to a vertical intermediate storage position to allow the wet film of varying heights to run off (Figure 7.32). In the meantime, the cleaned comb doctor blade is returned to its parking position for the next application. After a predetermined, sufficiently long-time window, the substrate is returned to the horizontal position and stored in the system's storage until the image analysis. Depending on the automation provider, however, an image analysis can be carried out before the film drying is complete in order to determine the limit value of the run-off. However, this is also possible at a later point in time in the dried coating film.

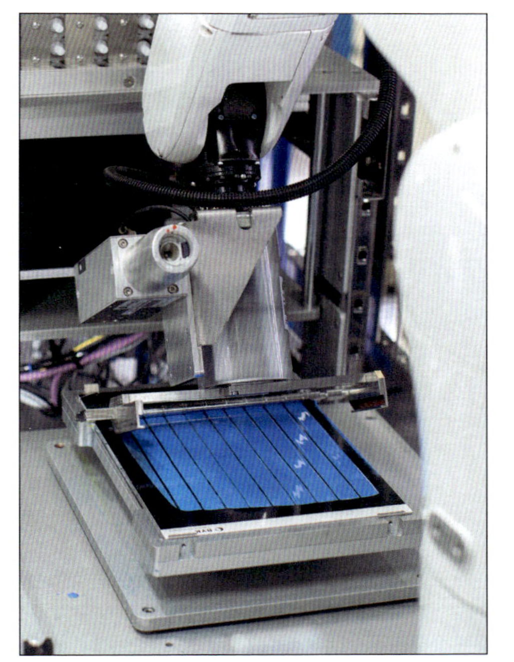

Figure 7.31: Automated comb doctor application with multi-axis robot Source: Chemspeed Technologies

7.2.2 Rub out

Performing the rub out test manually is a subjective method, as there is no standardized finger in humans and not every person performs the test with the same finger pressure, movement

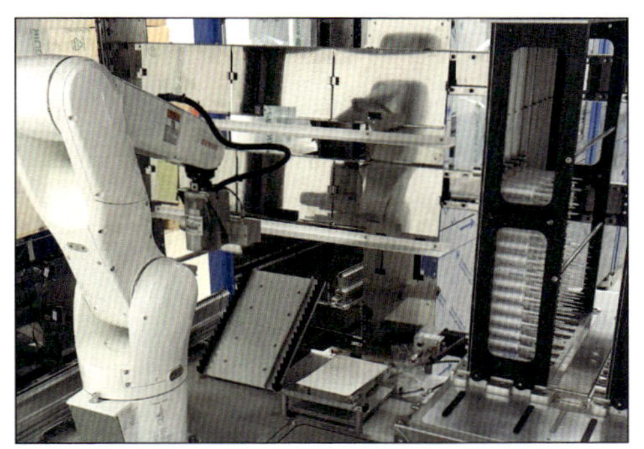

Figure 7.32: Automated sagging test module with intermediate storage Source: Chemspeed Technologies

and speed. In principle, this is not a tragedy, as it is more about a qualitative than a quantitative statement. However, for the assessment in the automated rub out version, the measurement with a spectrophotometer is necessary and therefore very objective [4]. The rub out test of the wet film is carried out after a predefined time after application, thus ensuring the comparability of different samples. Very often this happens immediately after application. The automation manufacturers use disposable consumables as rub out test bodies. The test specimens are provided singularly to the transfer position by feeding systems. From there, they are gripped by the robot using a vacuum. The robot arm moves the substrate with the wet film on the sample table. Depending on the manufacturer, the rub out movement is then carried out in a rotating or reciprocating manner (Figure 7.33). Füll Lab even enables the rub out test simultaneously performed on a parallel draw down (Figure 7.34). The movement parameters can be configured by the user. Depending on the supplier, the automated movement is carried out either via the robot arm with the test specimen or via the sample table with the substrate. What they have in common is that a uniform movement is always exerted with the same pressure, which is why the method is objective. At the end of the rub out test, the test body is disposed of in the discharge chute and the colour film is stored in the storage. After drying, any colour change is measured using a spectrophotometer. Depending on the programming of the workflow software, a pass/fail decision can mean the premature end of further planned tests on the sample.

7.2.3 Stippling

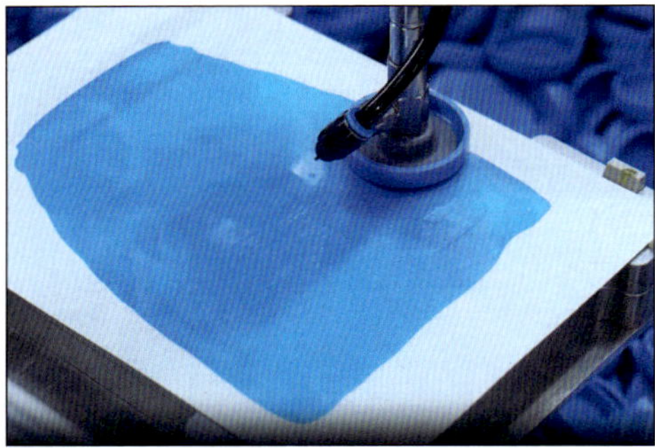

Figure 7.33: Fully automated rub out test performed on a single coating film
Source: Chemspeed Technologies

Figure 7.34: Automated rub out test on a parallel draw down setup
Source: Füll Lab Automation

A similar, but not comparable, behaviour to the sagging test is determined in the stippling test. This involves simulating the behaviour of an emulsion paint during roller application. Depending on the roller structure, force, speed and flow behaviour of the emulsion paint, a structure is visible in the wet film immediately after application. Depending on the flow properties of the emulsion paint, these are visibly reduced during film formation or disappear almost completely. In addition, air/foam bubbles are also left behind during application, which disappear depending on the defoaming. However, it is also possible that these bubbles burst too late, so that the bubble craters can no longer disappear due to the advanced film formation. Chemspeed has developed the stippling test method to objectively assess this behaviour. Immediately after the application of the coating film, the substrate is transferred to the sample table in the stippling module using a robot and fixed in place by vacuum. The behaviour of the stippling, analogous to the application of the real paint, can vary depending on the substrate. For this reason, substrates with differently smooth surfaces and absorption behaviour can also be used for testing. The robot's multi-gripper retrieves a sponge from the warehouse. The sponge is available in different soft or harder, stiffer, but always squeezable versions. The sponge is then repeatedly pressed onto the wet film in a predetermined pattern, leaving traces comparable to a roller application (Figure 7.35). At the end of the stippling process, the substrate is transferred to storage for drying. Depending on the paint, the surface structures achieved may remain visible or disappear after film formation. The result of the behaviour is determined after drying by means of optical image analysis. To do this, the substrate must be transferred from the warehouse to the imaging module and returned to the warehouse after the image analysis.

7.2.4 Pour out

Chemspeed offers the pour out test as a further method of assessing the behaviour of a paint or coating with regard to degassing or destabilization of foam. A minimum quantity of 50 ml of the wet sample is required to carry out the test. To prepare for the test, the multi-axis robot retrieves a large supporting sheet from the feeder and places it on the vacuum table. The robot then retrieves a transparent film with a maximum thickness of 1 mm from the second feeder and places it in the upper area of the vacuum table, slightly overlapping the supporting sheet. The vacuum table is then tilted to the desired or predefined angle. In the next step, the previously dispersed wet sample is gripped by the robot's gripper arm and guided over the inclined vacuum table. There, the cup is tilted for emptying and simultaneously guided along the upper edge of the transparent film (Figure 7.36 to Figure 7.37). This results in a film of paint running off the inclined transparent film. Depending on the degassing behaviour of the paint film, the air bubbles disappear to a greater or lesser extent. The remainder of the wet sample is emptied into the waste container provided for this purpose and the glass bottle is discarded into a separate waste container. An image is taken by the digital camera integrated opposite the tilted vacuum table. Digital image analysis is used to reg-

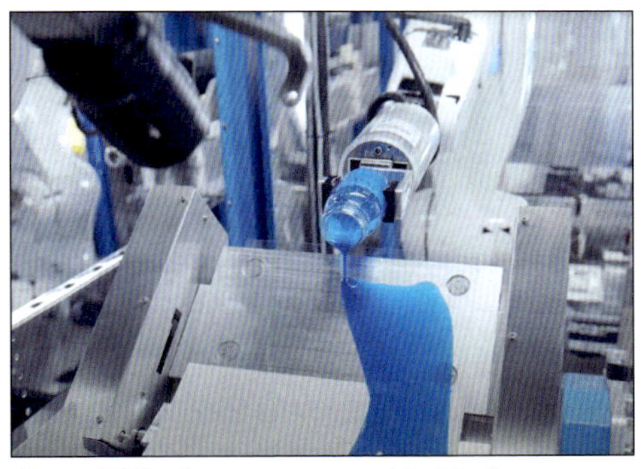

Figure 7.36: Pour out setup with tilted vacuum table and transparent sheet Source: Byk-Chemie

Figure 7.35: Fully automated stippling test integrated in the HTS of Byk-Chemie Source: Byk-Chemical

Figure 7.37: Automated pour out for analyzing defoaming behaviour Source: Byk-Chemie

ister the size and number of bubbles in the paint film and the results are evaluated using developed algorithms and finally transferred to the database. The film and the supporting sheet are then discarded in the waste container, just like the empty glass bottle.

7.2.5 Wet film imaging

Digital image analysis is also increasingly being used for the quantitative and qualitative evaluation of coating films that are still wet. It enables the measurement of geometric parameters (e.g. particle size, distance, area) and optical properties of the still-wet coating film. It is also suitable for the statistical analysis of defects, for example craters and pinholes, or material properties. Digital image analysis is also excellently suited for comparison with reference values and measurement results with defined tolerance ranges or standards [5, 6]. Automated systems for wet film images can be found in the portfolio of Labman and Füll Lab. Labman offers a flexible three module system capable of producing and testing development samples of coatings. Using liquid raw materials, the preparation module can produce over 100 samples in a day with orders of addition controlled by the user. These samples can be moved to the application module for application of the coatings to substrates with "Byk Bird Bar" film applicators. Wet colour and opacity information can be collected with a non-contact spectrophotometer (Figure 7.38). This module also includes instruments for collecting tinter compatibility, rheology and film thickness data. The application module creates 400 films during a day of operation. After curing, the applied films can be further studied with the analysis module. This module has an industry standard contact spectrophotometer, a balance and a laser profilometer integrated to provide film weight, thickness and optical data. This module analyses 400 applied films in a day.

7.2.6 Film drying

Figure 7.38: Determination of colour and opacity of wet coating films
Source: Labman Automation

The drying recorder is an in-house development from Füll Lab. The measurement is carried out in the same way as standard drying recorders. Coated substrates are placed in the module directly after coating, a test syringe is automatically lowered and moved over the sample at a defined speed. At the end of the test, a photo is taken of the sample. The evaluation can be carried out using image processing. The standard module can test up to 50 samples, which are produced one after the other on the system, in parallel.

7.3 Characterization of dry coating film

Since paints and coatings have to meet various requirements such as aesthetics, signalling, protection against external influences such as high-energy radiation or corrosive media, but also mechanical resistance, the number of tests on dried or cured paint films is undoubtedly very large. The tests can be divided into optical, mechanical and chemical tests as well as destructive and non-destructive methods. Digital image analysis is a modern method that is becoming increasingly important in the paint and coatings industry. It is used to systematically extract information from images in order to evaluate various properties of coatings, paints and varnishes. Specialized software is used to analyze image data quantitatively and qualitatively. For paints and coatings, it is used to examine the quality and structure of the coated surfaces but does not exclude the use of wet coating films. It is mostly used on coating films that have already dried or hardened, for example to check the thickness of the coating layer and its homogeneity, assess micro- and nanostructures, analyze the distribution of pigments and fillers, detect and quantify corrosion spots after salt spray tests or analyze cracking or peeling after weathering tests. The advantages are obvious:

- Objectivity: Automated evaluation eliminates subjective errors
- Precision: High accuracy in the measurement of micro- and nanostructures
- Time saving: Faster analysis compared to manual methods
- Reproducibility: Results can be standardized and repeated
- Versatility: Applicable to a variety of parameters (colour, gloss, structure, etc.)

7.3.1 Optical coating characterization

The optical properties of all paints and coatings produced are actually determined without exception. Customers often remain very loyal to their suppliers, especially in colorimetry, which is why laboratory automation providers also have to flexibly integrate different manufacturers. Along with colorimetry comes the testing of reflective properties with different measuring geometries and the gloss haze. Of course, these properties are also subject to the coating thickness, which is why this is also usually measured. Due to the correlations listed above, the measurement methods are usually concentrated in one module. This also makes sense due to the comparable substrate handling. Depending on the manufacturer, different methods are used for the logistics of the substrates in the automation of different processes. In contrast, vacuum grippers or vacuum plates are primarily used to characterize the optical properties. This allows the substrates, which are not always rigid, to be placed flat against the measuring instruments and reliably deliver reproducible results. A few examples of spectrophotometers in laboratory automation are presented here, which can be transferred to all other manufacturers. However, instruments from Datacolor, X-Rite, Byk-Gardner and Minolta are integrated most frequently (Figure 7.39 to Figure 7.42). The arrangement or selection of the instruments is based on the substrate feed from the underside. This

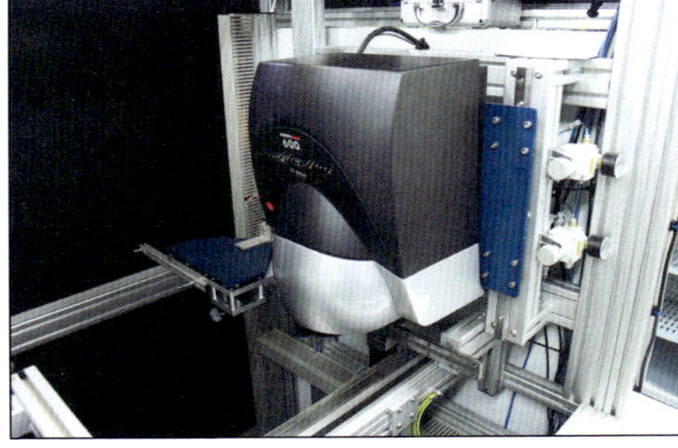

Figure 7.39: Automated colour measurement with Datacolor spectrophotometer

Source: Labman Automation

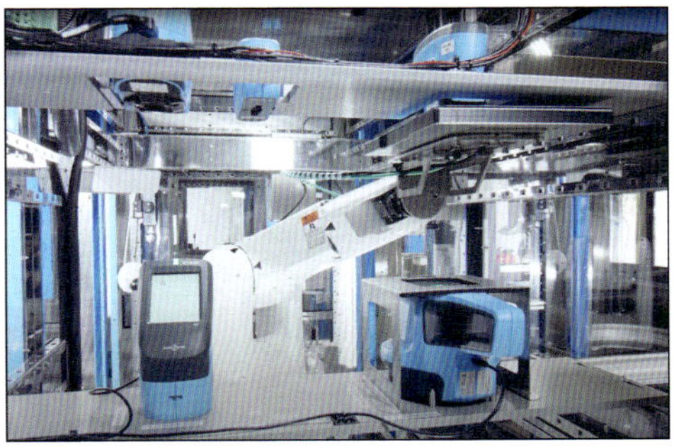

Figure 7.40: Coating characterization module with various optical instruments

Source: Byk-Chemie

applies analogously to the measurement of gloss, gloss haze and dry film thickness. The coated substrate is transferred from the storage to the characterization module. Within the module, the integrated multi-axis robot takes over the transfer to the individual measuring instruments with a vacuum plate. Depending on the requirements and substrate, several determinations are made for brightness, colour values, gloss and gloss haze as well as coating thickness at different points. To determine the opacity of a colour film, measurements are of course taken over a white and black background and the result is calculated. Once all measurements have been completed, the coated substrate is transported back to the transfer station and back to the warehouse. Optionally, measurements of the surface quality can also be realized by means of wavescan.

The "Q-Chain Surface Scanner Automatic" from Orontec is a multi-automatic measuring machine with which flat test specimens (usually painted sheets or panels) can be measured in a predefined measuring grid using various standard measuring devices and statistically evaluated using integrated software [7, 8]. The standard version of the measuring devices used are hand-held

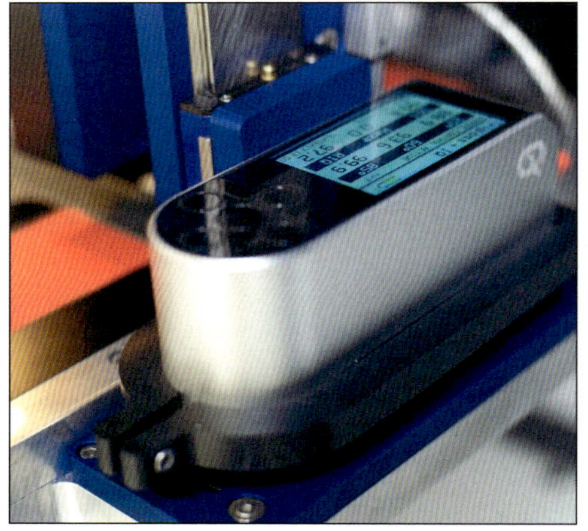

Figure 7.41: Integrated handheld gloss device in automation setup Source: Labman Automation

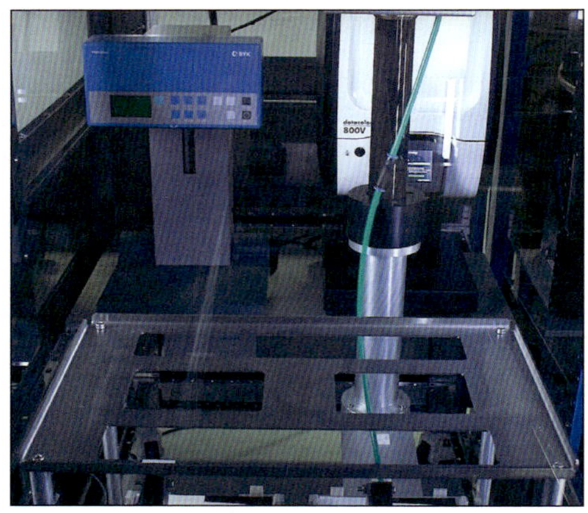

Figure 7.42: Side by side integration of colour and tri-gloss measurement Source: Omya

Figure 7.43: "Q-Chain Surfcace Scanner Automatic" of Orontec Source: Orontec

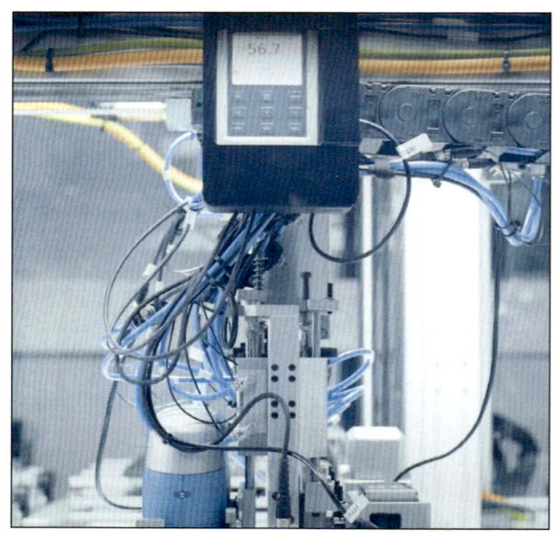

Figure 7.44: Measuring of dry film thickness in "Q-Chain Surface Scanner Automatic" Source: BASF

measuring devices that can be triggered externally and the measured values read out directly by integrating them into the automatic measuring machine. Test specimens can be measured and evaluated for pure colour measurements (e.g. colour samples) or for pure coating thickness measurements (e.g. spray pattern analyses). The possibility of combining several measuring devices also allows statements to be made, e.g. on the correlation between coating thickness, colour and surface structure. This means that it can also be used for evaluations in accordance with ISO 28199. The standard measuring devices are a magnetic inductive coating thickness gauge,"Byk-mac i colorimeter", "Byk Wave-scan dual" and "Byk cloud-runner" (Figure 7.43 to Figure 7.45). The "Byk Tri Micro Gloss µ" and the "OptiSense Paint-Checker Automation" can be integrated as an option.

Figure 7.45: Integrated "Byk wave-scan dual" in "Q-Chain Surface Scanner Automatic" Source: Orontec

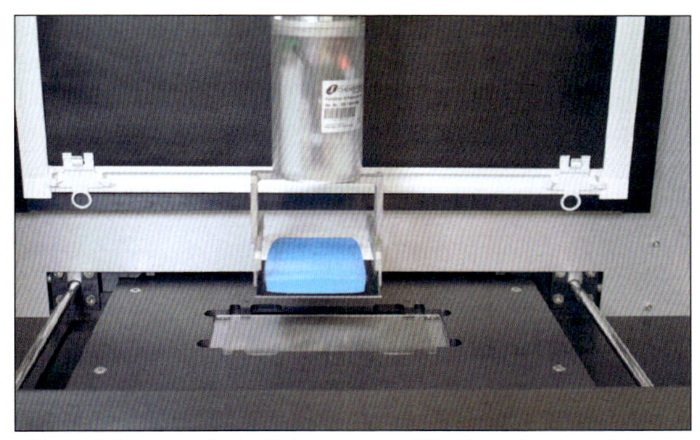

Figure 7.46: Loading of dry coating film into the imaging drawer Source: Chemspeed Technologies

7.3.2 Film imaging

The digital image analysis of coating surfaces is an enormous help in the evaluation and optimization of paints, inks and coatings. High-resolution images or microscopic images are analyzed to obtain information about the structure, quality and defects of surfaces. It is also possible to obtain "only" the measurement of colour data and the surface quality and thus offers the best prerequisite for laboratory automation for qualitative and quantitative evaluation. For the neutral, comparable evaluation of the coated, dry films on the substrates are delivered to the imaging module. The substrate is loaded into the extended

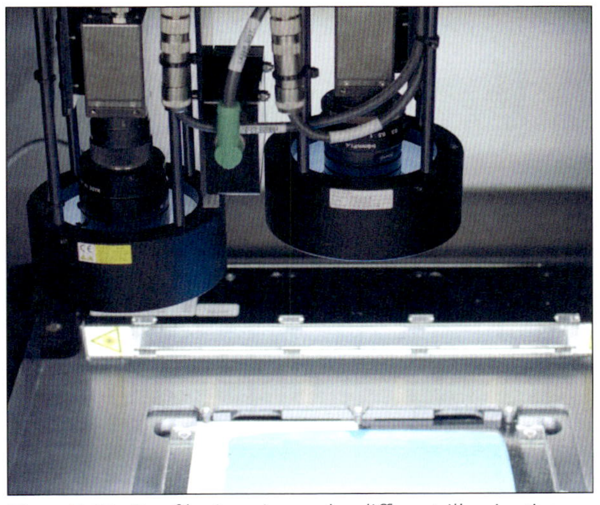

Figure 7.47: Dry film imaging under different illumination Source: Chemspeed Technologies

drawer of the darkroom using a linear overhead robot (Figure 7.46). The substrate is then moved into position with the retracting drawer. The substrate is held flat with the aid of a vacuum holder. The darkroom can be equipped with one or more camera sets (Figure 7.47). There is also an integrated motorized lighting unit which enables various exposure conditions. In addition, there are lateral exposure units so that total exposure from four directions is possible. When using transparent substrates, illumination can also be carried out from below the substrate. Furthermore, there is a height adjustment of the substrate to realize different exposure angles. Depending on the setup, different tasks can be realized with the imaging module. One example is the evaluation of the quality level of the cross-cut inspection. The evaluation is carried out using an algorithm developed for this purpose. The results are written to the workflow software for further use.

7.3.3 Tack & stickiness

There are various methods for determining the surface tackiness of coating films, some of which are very subjective using human touch (thumbs), but there are also very objective methods. These are particularly suitable for automation in order to determine influences caused by drying time, temperature, humidity, coating composition or coating thickness, for example. As surface tack is very often associated with the drying or through-drying of paint films, one of the requirements is to determine this at time intervals. The substrate is delivered to the measuring module in accordance with defined specifications regarding the start of the measurement and placed on the measuring table by the Multi-gripper tool on the robot. For the measurement, the substrate is fixed to the measuring table using a vacuum setup. After changing the tool from the gripper to the measuring tool with fully integrated force gauge, the measurement can begin (Figure 7.48 to Figure 7.49). In addition to a single measurement, interval measurements can also be carried out at defined intervals of at least two minutes. Between each

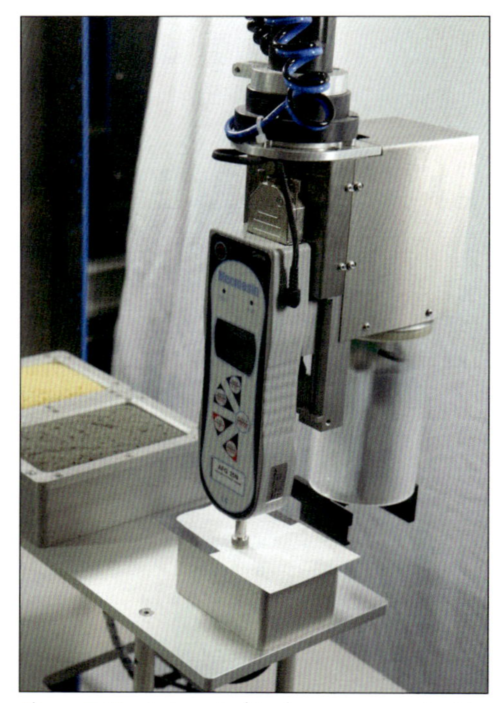

Figure 7.48: Automated tack measurement with "Mecmesin" force and torque indicator
Source: Chemspeed Technologies

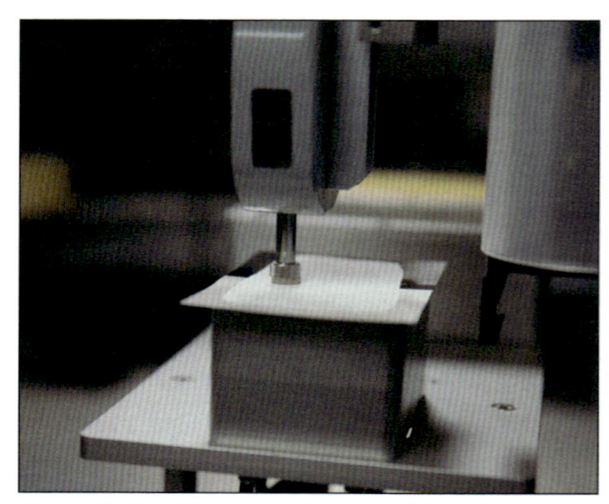

Figure 7.49: Measuring head positioned on the coating film
Source: Chemspeed Technologies

measurement, the measurement body is cleaned with the integrated sponge system to eliminate influences from the previous measurement. The measuring accuracy is ±0.15 %. After the measurement, the substrates are either discarded in a waste container or transported back to storage.

7.3.4 Hardness

The pendulum hardness test is a proven method for evaluating the surface hardness and elasticity of coatings. The pendulum hardness test is based on the oscillating movement of a defined pendulum on a coating surface. The hardness is measured by the number of oscillations or the time it takes for the pendulum to stop oscillating due to energy loss. Depending on requirements, either the pendulum hardness according to König (DIN EN ISO 1522) or the pendulum hardness according to Persoz, which is more sensitive for softer coatings and has a higher degree for the start position of oscillation, can be configured. In addition to many other mechanical tests, Füll Lab also offers an automated solution for measuring pendulum hardness. The substrates are inserted into the pendulum hardness tester by the robot's substrate tool. Once the pendulum's damping balls have been placed on the coating film (Figure 7.50), the robot arm's double tool deflects it to the corresponding angle of 6° or 12° and starts the measurement by releasing the holder pin (Figure 7.51). After damping to 3° or 4°, the measurement is ended, and the result is transferred to the database. The

Figure 7.50: Coated substrate under oscillation of the pendulum
Source: Byk-Gardner

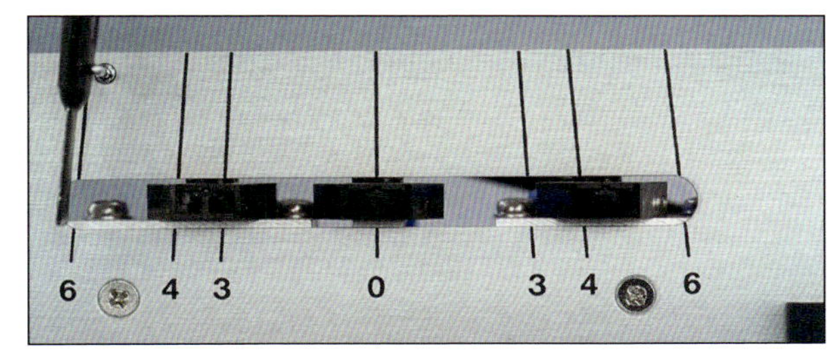

Figure 7.51: Pendulum locked by the pin before release
Source: Byk-Gardner

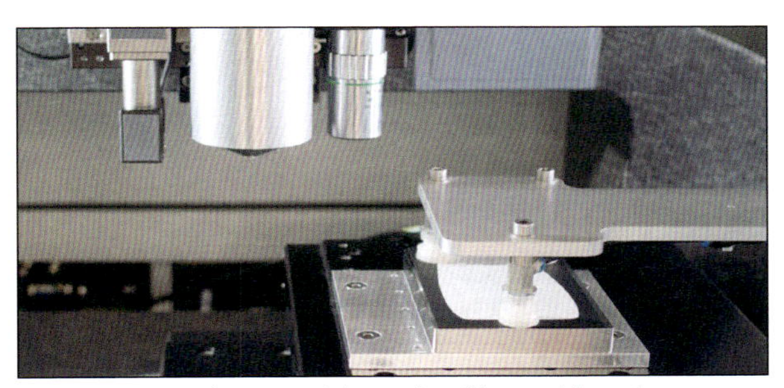

Figure 7.52: Substrate with coating film positioned on measuring table
Source: Chemspeed Technologies

Figure 7.53: Measuring head with Berkovich diamond before test launch
Source: BASF

substrate is lowered again and then removed by the robot arm and returned to the original recording position.

7.3.5 Nano indentation

Nano hardness is a precise method for measuring the mechanical properties of materials in the nanometer range. Nano hardness is measured by the penetration of a hard test specimen (e.g. Berkovich diamond) into the coating or polymer film under a defined force. The penetration depth under load and unload is recorded in order to determine the mechanical properties. The fully integrated automatic module is positioned on an anti-vibration air table with the "Step 500 nanoindenter" from Anton Paar and enables nanomechanical testing and analysis of the surface topography on flat substrates. The configuration consists of the nanoindentation head (NHT3) with Berkovich diamond indenter from 0 to 500 mN and a microscope with 20x and 100x micro-objectives. The platform is equipped with a multi-axis robot and multi-gripper for fully automatic operation of the module. The multi-gripper is used to manoeuvre the substrate to be measured from the transfer station onto the sample table and place it there (Figure 7.52). After approval by the sensor system, the measurement is started, and the results are written to the run execution software at the end (Figure 7.53). Finally, the sample is removed with the multi-gripper and returned to the transfer station. Fully automated solutions have so far been offered by Chemspeed Technologies and Füll Lab.

Figure 7.54: Dispensing of wash solution onto coating film
Source: Chemspeed Technologies

7.3.6 Wet scrub resistance

Wet abrasion resistance describes the resistance of a coating to mechanical stress caused by the movement of a brush or an abrasive scouring pad when wet. Two different standards from ISO or ASTM can be used for this purpose, although they differ significantly. The developed method of Chemspeed presented below is not standardized but is based on it and is therefore suitable for screening emulsion paints. The fully automated module consists of several elements. A weighing device with adjacent heating plates is responsible for weighing the substrate before and after the test. The test itself takes place in a closable

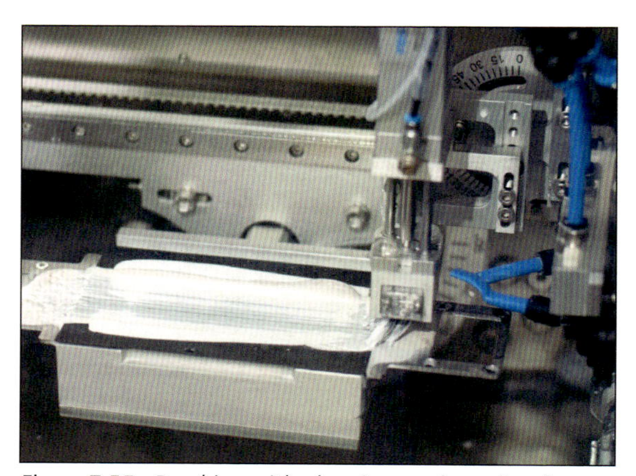

Figure 7.55: Brushing with abrasive scrub media of coating film
Source: Chemspeed Technologies

scrubbing chamber and the cleaning and drying takes place in the second adjacent chamber. The workflow starts with the transfer of the coated PVC film to the weighing station to determine the unscrubbed substrate weight. The gripper of the linear axis overhead robot then picks up the substrate and transfers it to the vacuum table in the open scrubbing chamber before it is closed. Depending on the selected workflow setup, either the sodium polyphosphate wash solution or the abrasive ASTM scrubbing paste is applied (Figure 7.54). The scrubbing process then starts with a brush and a defined number of scrubbing cycles (Figure 7.55). At the end of the scrubbing cycle, the scrubbing chamber opens, and the robot picks up the substrate and transfers it to the washing and drying chamber on a vacuum table. The chamber is closed and the vacuum table with the sample is tilted to a 45° position. The residues from the scrubbing process are washed off with nozzles moving in parallel over the substrate surface. Finally, the scrubbed paint film and the substrate are blown dry. The chamber then opens, and the robot transfers the substrate between the heating plates to remove any remaining water droplets. As the penultimate step, the substrate is returned to the weighing device to determine the weight loss caused by the scrubbing process. The workflow ends with the robotic transfer of the substrate from the weighing device to the starting position for further transfer to the substrate store and labelling.

Figure 7.56: Burnish resistance module with storage of polishing bodies Source: Chemspeed Technologies

7.3.7 Burnish resistance

Burnish resistance refers to the ability of a paint or coating to resist the formation of shiny marks or gloss changes on its surface when subjected to rubbing, brushing, or light abrasion. This property is particularly important for interior wall paints and coatings used in areas prone to frequent contact or cleaning. Automation manufacturers have this test developed since years, e.g. Labman and Chemspeed. The burnish resistance module is equipped with an overhead linear axis robot for logistics. This picks up the PVC substrate with the dried paint film from the transfer station and places it on the vacuum table of the burnish station. One of the 60 burnish bodies is then retrieved from the warehouse and

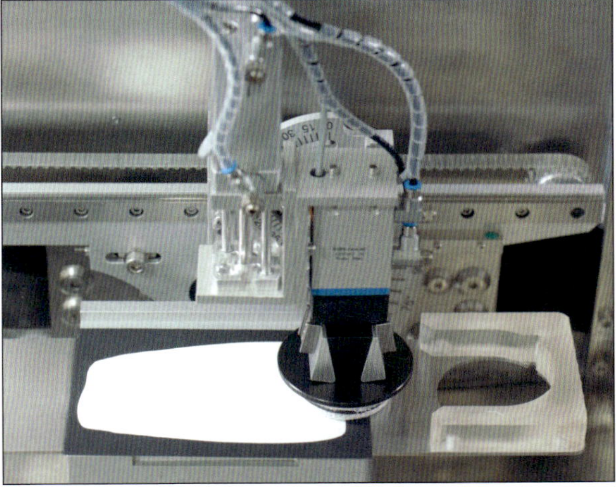

Figure 7.57: Polishing of coating film to determine burnish resistance Source: Chemspeed Technologies

placed in the burnish station (Figure 7.56). The burnish carriage picks up the burnish body and performs a defined number of polishing cycles. The number of cycles and the speed of the polishing process can of course be parameterized (Figure 7.57). At the end of the cycle, the burnish body is returned to storage, and the substrate is placed back in the transfer station. The resistance of the paint film to polishing and the associated increase in gloss is then determined in the further workflow in the characteristic module. There the polishing is determined with a gloss meter under certain angle geometries, most relevant at 85°, and the results are transferred to the database.

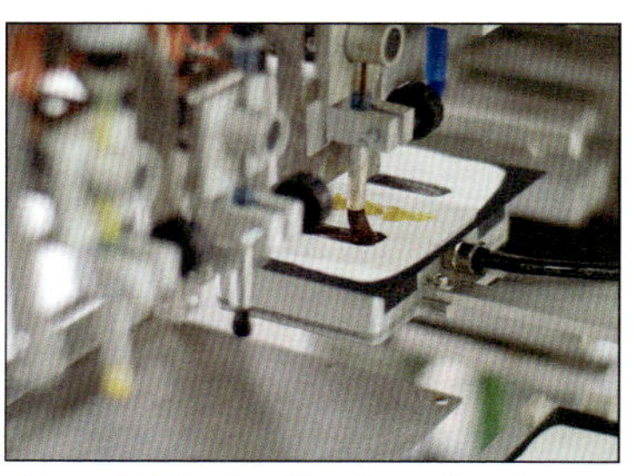

Figure 7.58: Application of stains onto the coating film
Source: Chemspeed Technologies

7.3.8 Stain resistance

Stain resistance refers to a coating's ability to resist staining from various substances, such as dirt, oils, food, beverages, and chemicals, without absorbing or discolouring. It is a critical property for paints and coatings used in areas prone to frequent spills, contact with hands, or exposure to contaminants. Standard test methods such as ASTM D4828 and ISO 2812-4 describing the execution of the test. Different stains such as coffee, red wine, ketchup, mustard, lipstick, pen marker, etc. can be used for the determination of the stain resistance. A stained coated panel is subjected to cleaning with a standard cleaning solution. The effectiveness of cleaning is assessed by comparing the surface before and after cleaning. The workflow of the fully automated module for determining stain resistance begins with the transfer of the PVC substrate to the application area. The overhead linear axis robot guides the substrate with the coating film under the corresponding 240 ml cartridge with the stain to be applied. During dispensing, the stain is simultaneously distributed with a brush by the substrate movement using the robot arm (Figure 7.58). Marking with a permanent-coloured pen is also possible. Once the stain has been applied, the substrate is transferred to the rinsing station. The excess amount applied is rinsed off and then the substrate is temporarily stored for the stains to take effect. After a defined exposure time, the robot transfers the substrate to the vacuum table of the rinsing station. This is closed and the cleaning carriage picks up the sponge that has already been transferred from the bearing. The sponge, conforming to ASTM or ISO, is moved back and forth over the coating film with simultaneous dosing of the cleaning agent (Figure 7.59). The number of strokes and speed can be freely defined. After the washing process, the used sponge is thrown into a waste container. The substrate with the cleaning solution and more or less washed off stains is

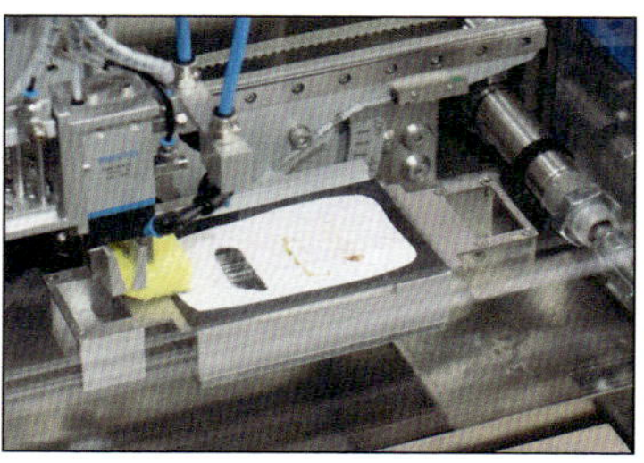

Figure 7.59: Cleaning of stains with sponge strokes over the coating film
Source: Chemspeed Technologies

Figure 7.60: Electrochemical corrosion assessment system
Source: Labman Automation

then transferred to the rinsing station. After placement on the vacuum table, it is tilted back to a 45° position. The rinsing process begins and is dried by the drying process with the air knife. Automated solutions for the stain resistance test are offered by Chemspeed as well as Labman.

7.3.9 Chemical resistance

Chemical resistance refers to a coating's ability to withstand exposure to various chemicals – such as acids, alkalis, solvents, and oils – without degrading, discolouring, or losing adhesion. It is a crucial property for coatings used in industrial, marine, and protective applications where chemical exposure is frequent. The demand for this test is relatively frequent, which is probably why the three suppliers Chemspeed, Füll Lab and Labman have included this test method automatically in their portfolio. The electrochemical corrosion assessment system of Labman performs electrochemical tests on small samples of metal (Figure 7.60). The system consists of a central robot arm with a bank of electrochemical testing modules on each side. Samples are loaded into cassettes at one end of the system before being moved into the modules by the robot arm (Figure 7.61). The modules seal onto the samples, then fill their cells with test solution (Figure 7.62). A variety of different tests can be performed using the module's potentiostat. After the test has been completed the sample and cell are washed and dried before being returned to the cassettes by the arm. Currently there are 10 modules in the system, with space to expand to 16 in the future. All modules can be run in parallel, resulting in a high throughput. There is capacity for 132 samples in the system, in easy to change cassettes. Optionally, pairs of cells can be joined together to test two samples for galvanic corrosion. The size of the interface between the test cell and the sample can be quickly changed to various pre-set sizes. Test cells can be aerated or de-aerated depending on the test required.

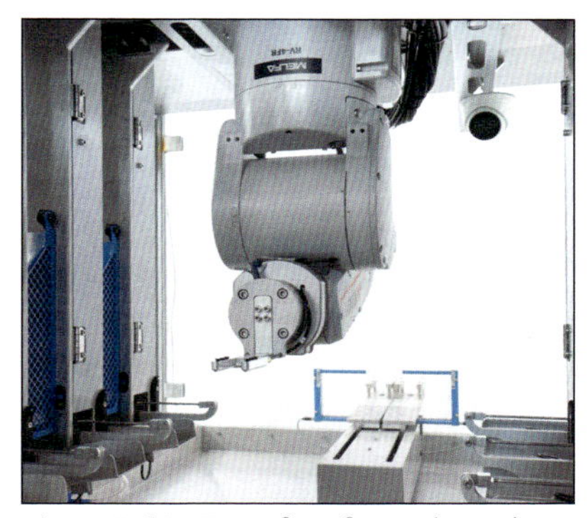

Figure 7.61: Transfer of sample to the test cell Source: Labman Automation

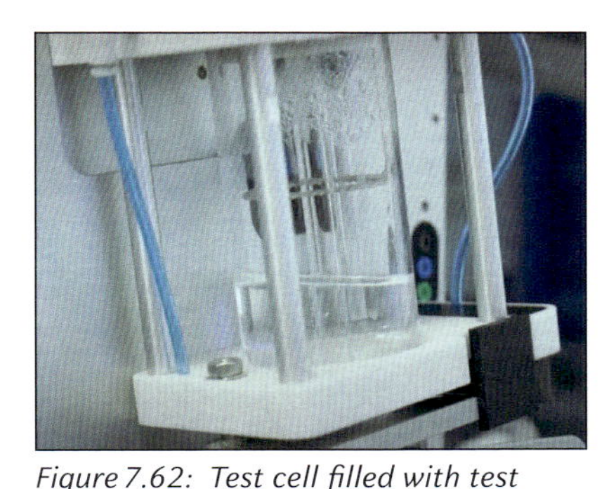

Figure 7.62: Test cell filled with test solution and sample below Source: Labman Automation

7.3.10 Corrosion assessment

For the fully automatic measurement of corrosion protection sheets Orontec has developed the "Q-Chain SuMo" (Figure 7.63). The evaluation of corrosion protection test sheets is often carried out based on DIN EN ISO 4628 – 8 using visual assessment with the aid of a ruler or similar measuring equipment. This is both time-consuming and subject to errors. The "Q-Chain

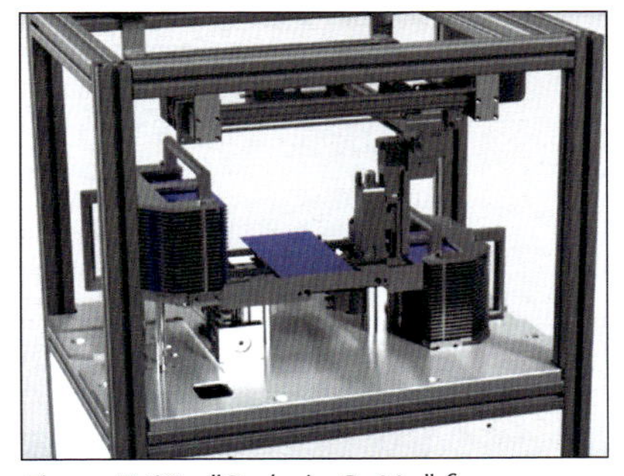

Figure 7.63: "Q-chain SuMo" for automated corrosion assessment Source: Orontec

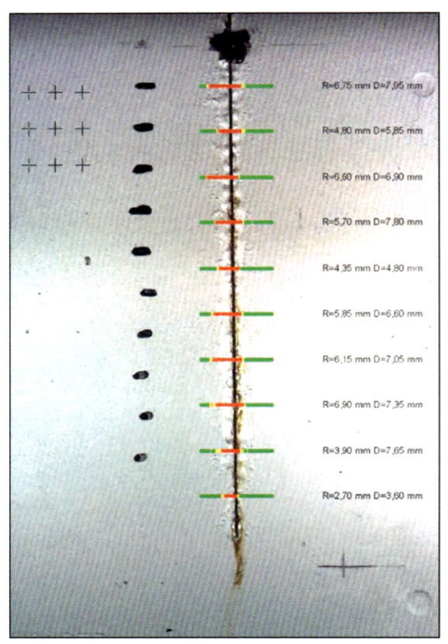

Figure 7.64: Corrosion assessment by pulse thermography *Source: Orontec*

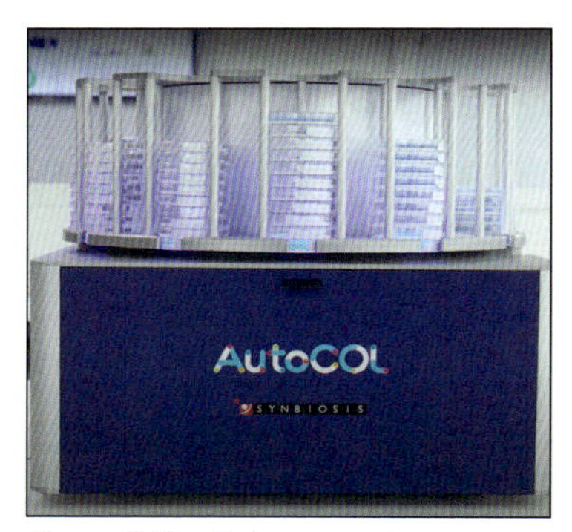

Figure 7.65: Colony counter "AutoCOL" for automated de-lidding and imaging of petri dishes
Source: Labman Automation

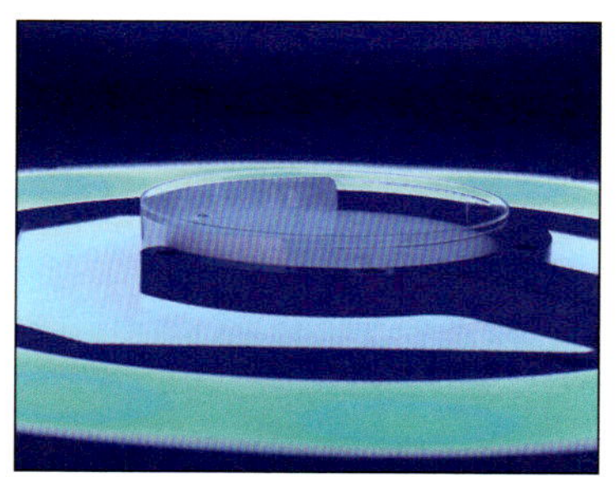

Figure 7.66: Imaging cell with loaded petri dish for inspection
Source: Labman Automation

SuMo" is a fully automatic device for measuring corrosion protection test sheets that covers this process from the creation of measurement grids to the evaluation. The sensors used also eliminate the need to remove delaminated or rusted areas and they are based on pulse thermography, which can be used to differentiate between delamination and corrosion (Figure 7.64). With this measuring method, measurement signals differ significantly between adhesive and non-adhesive layers, so that very good differentiation is possible. The sensors are integrated into an XY machine (approx. 60x50x50 cm) that enables rasterization with a resolution of 0.15 mm. The method does not require the removal of the detached paint material ("scraping") from the test object and on the one hand saves the effort of processing the test objects, and on the other hand it reduces the user influence due to mechanical effects on the coating. The method also potentially opens up the possibility of using test specimens' multiple times. The test sheets can be further weathered after measurement and a time series can be measured using just one sheet. The device has an integrated device that allows the sheets to be processed in batches. This enables the testing process to be automated, even for a series of test specimens. The sheets are usually marked using barcodes or QR codes that are read by an integrated camera. This also automates the assignment of the results to the test specimens in the database. Evaluations, for example according to DIN EN ISO 4628 – 8, contain not only information about the average width but also a visualization of the measurement data. Here, for example, the areas that are delaminated (yellow) and rusted underneath (red) can also be visualized. Because the sheets do not have to be reworked ("scraped"), multiple weathering tests can be carried out on one and the same test specimen. Differentiating between delamination and rust underneath can also provide further insights.

7.3.11 Microbiology

A bespoke product for automating the de-lidding and imaging of petri dishes. Labman presents a revolutionary automated colony counting system (Figure 7.65 and Figure 7.66 . Developed in partnership with Synbiosis, it utilizes a carousel that holds one hundred 90 mm petri dishes. The productivity is increased by 700 % compared with manual operation. It de-lids plates from the carousel stacks, reads their barcode, takes an image of the dish's contents. Then it reunites the bottom half of the dish with its lid and places it back into the carousel. The carousel stacks are illuminated using RGB LED ring lights to allow quick visual status updates. A full run of 100 plates takes around 40 min to complete. Before each run begins the software allows the operator to customize all aspects of the imaging and light settings used to take the image. Barcode information can also be pre-loaded into the software.

7.3.12 Various other dry coating characterizations

The number of test methods already presented is surprisingly high, but still not complete. Of course, there are more and the demand for further, especially subjective and time-consuming, methods will continue to grow. Finally, further methods for characterising dry, cured coating films are listed alphabetically without evaluation by automation providers.

  - Chemspeed Technologies
  - Abrasion resistance
  - Multitesting for friction and wear
  - Cross cut
  - Ellipsometry
  - X-ray photoelectron spectroscopy
  - X-ray diffraction
  - Scanning electron microscopy
  - Dynamic mechanical analysis (DMA)
  - FT-MIR (ATR element)
- Füll Lab Automation
  - Conductivity
  - Pot life
  - Crosscut
  - Mandrel bending
  - Scratch test
  - Erichsen hardness
  - Static tensile shear test
- Labman Automation
  - Laser profilometer
  - Semi-automated abrasion resistance testing

7.4 Literature

[1] DIETRICH, R.; "Paint Analyis"; 2nd Edition; ISBN 9783748604259; Vincentz Network, Hannover; 2021

[2] KETTLER, W., et al.; "Colour Technology of Coatings"; ISBN 9783866306998; Vincentz Network, Hannover; 2016

[3] SCHULZ, U.; "Accelerated Testing", ISBN 9783866309081; Vincentz Network, Hannover, 2008

[4] BRAMLAGE, C.; REUTER, E.; "Automated on the overtake track", p. 42ff, Farbe und Lack, Vincentz Network, 4/2022

[5] HUSTERT, H.; WÖRHEIDE, R. J.; "Präzise Flüssiglackfarbmessung", p. 48ff, Farbe und Lack, Vincentz Network, 3/2014

[6] HUSTERT, H.; WÖRHEIDE, R. J.; "Grüne Welle – vom Lackdesign zur Serienproduktion", p. 150ff, Farbe und Lack, Vincentz Network, 4/2017

[7] Orontec; "Q-chain Surface Scanner Automatic", Technical Description, 2024

[8] BASF, "Resins for Coatings Applications: Speeding up innovation with digital material profiling", https://youtu.be/S6CIuP4wv_U?si=wfiAuGtvHcoQGdPG, 24.03.2021

8 Digital solutions

Digital solutions refer to the use of digital technologies, software, and systems to address specific problems, enhance processes, and achieve business or operational goals. They are integral to modern industries, enabling innovation, efficiency, and scalability. Research and development are in the process of being transformed towards automation and digitalization. The range of possibilities is increasing, as is the desire for high sample throughput as well as standardization within companies with a presence in different countries or even continents. But what is the difference between digitization and digitalization? Digitization can be described as any time you translate something into bits and bytes – for example, by scanning a photo or a document – you are digitizing that object. The next logic step comes with digitalization, when data from throughout the organization and its assets is processed through advanced digital technologies, which leads to fundamental changes in business processes that can result in new business models, standard operations, social change, and many more. Supported by automation and digital technologies, research & development (R&D) can unlock new value from data that has been aggregated through digitization, drive innovation, and allows faster time to market. These are the benefits of digitalization (Figure 8.1). The overall effect of digitalization across an organization and thus in R&D is called digital transformation – more of a process than an outcome. Digital twins are disruptive for every lab and enable the artificial intelligence/machine learning (AI/ML) pipelines in the future combined with ready-to-apply and quality assured raw materials, intermediates, formulations, and their applications and characterization.

The key components of digital solutions are software applications, cloud computing, artificial intelligence (AI), machine learning (ML), internet of things (IoT), big data & analytics and blockchain.

- Custom or off-the-shelf tools for various functions (e.g., ERP, CRM, LIMS).
- Platforms like Amazon Web Services (AWS), Azure, or Google Cloud for scalable storage and computing power.
- Technologies for predictive analytics, automation, and data insights.
- Devices interconnected to collect and share data.
- Tools to process and analyze large datasets for actionable insights.
- Distributed ledger technology for secure and transparent transactions.

Applications of digital solutions in laboratory and research are laboratory information management systems (LIMS) for sample tracking and data management. This can be paired with automation and robotics for streamlining repetitive tasks like sample preparation. When having collected large datasets by HTE, data analysis tools such as AI can be used for processing these. For the implementation of digital solutions, it is advisable to consider the following aspects.

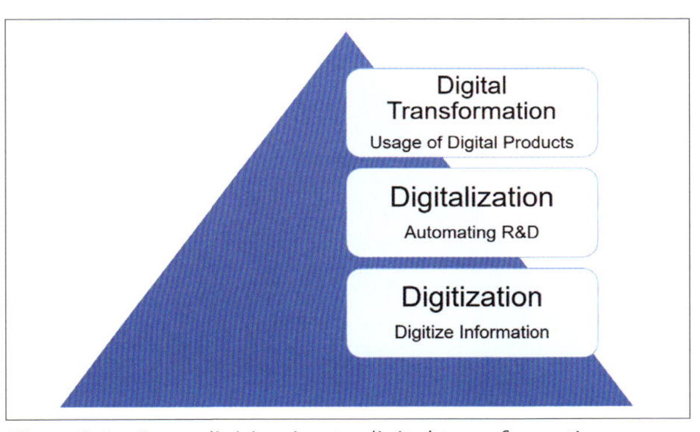

Figure 8.1: From digitization to digital transformation
Source: PERFECO Consulting Gysau

  - Identify needs by assess pain points and define objectives
  - Choose the right technology by evaluating tools and platforms that align with goals
  - Pilot projects are recommended to test solutions on a small scale before full deployment
  - Stakeholder buy-in is a requirement to have teams engaged and provide training to ease adoption
  - Monitor and optimize continuously performance and make improvements

8.1 User interface

User interfaces (UIs) are critical for delivering a positive user experience (UX). A well-designed UI ensures users can interact with a product, system, or service in an intuitive, efficient, and enjoyable way. By focusing on UI design principles and best practices, businesses can significantly enhance the overall UX, attitude and commitment of the users. Customizable and user-friendly platforms also enhance engagement of the users.

The core principles of UI Design for better UX are clarity, consistency, accessibility, feedback and responsiveness, simplicity plus user control. The use of clear labels, simple icons, and unambiguous navigation provides clarity. Ensuring users can easily understand the interface by minimizing the cognitive load. Very important is to maintain uniformity in design elements such as colours, fonts, and button styles as well as ensure consistency across all platforms (desktop, mobile, tablet). Mandatory are designed interfaces that accommodate users with diverse abilities (e.g., screen readers, high-contrast modes). The accessibility needs to follow guidelines like WCAG (web content accessibility guidelines). Furthermore, the provision of immediate feedback for user actions (e.g., loading indicators, confirmation messages) is essential. This requires responsiveness in both design (adaptive layouts) and performance (fast loading times). However, UI features should not overboard, rather kept simple. Avoiding clutter, instead prioritizing essential elements and minimizing distractions is key. The use of whitespace effectively to create a clean and organized appearance support the simplicity. And finally, give users control and allow them to undo actions or revert to previous states. Provide clear exit options or "cancel" buttons in workflows reduces frustrations.

Several of the below mentioned components enhance UX and should be mandatory elements in today's UI solution offers.

- Navigation
  - Menus ensure intuitive navigation through logical menu structures
  - Breadcrumb help users understand their location within the app or website
  - Include smart search features like autocomplete and filters
- Forms
  - Simple layouts with clear labels and real-time validation
  - Minimizing required fields to reduce user effort
- Buttons and icons
  - Design buttons with clear labels and distinguishable styles (e.g., primary vs. secondary actions)
  - Use of familiar icons that align with user expectations (e.g., a magnifying glass for search)
- Typography and colour
  - Legible fonts and appropriate font sizes for readability
  - Use of colour schemes that guide attention while maintaining visual harmony
- Interactive elements
  - Use of animations and transitions to make interactions more engaging
  - Including hover effects or tooltips for additional information

Modern UI practices enable tailored interfaces to user preferences. Mobile optimization ensure designs are mobile first and responsive to different screen sizes. Also, many modern dashboards include touch-friendly elements and gesture for mobile users. Users demanding the simplification of complex tasks by breaking them into smaller steps (e.g., progress wizards) as well as the use AI-powered assistance (e.g., chatbots). The scientific purpose does not exclude the incorporation of game-like elements such as progress bars, badges, or leaderboards to boost engagement, especially popular with younger generations. On the other hand, UI designs need to fulfil diverse user needs across demographics and devices. Finally, technology need to provide compatibility with evolving technologies such as augmented reality (AR)/virtual reality (VR), voice interfaces, etc.

8.2 Software solutions in automation

Run execution software refers to specialized platforms or tools designed to manage, monitor, and execute tasks, processes, or experiments within laboratories. These tools ensure consistency, efficiency, and traceability, often forming a critical component of digital transformation initiatives. Typically, the laboratory automation suppliers develop their own software solution to have full functional control and adaptability to their hardware solutions.

The single module run execution software of Chemspeed called "Autosuite" was designed to bridge the gap when transitioning a laboratory bench process to one of their automated workflow modules. Transforming a bench procedure to an automated workflow is challenging enough without the added complexity of translating human tasks into robotic instruction. They claim to have designed an interface that makes their automated systems speak the scientist's language to help ease this process. "Autosuite" provides an intuitive four stage process that helps unlock the full potential of Chemspeed's platforms while providing perfect balance between instrument control and experiment requirements. The feature list of "Autosuite" contains:

- Intuitive workflow: Template driven simplified programming of highly sophisticated automated systems
- Quick onboarding: Increase "right-first-time" runs on fully automated platforms, regardless of experience level of operators
- Full control: Enables users to unlock the full automating power by providing deep control of tools and functions
- Know what's happening: Enables full transparency of run execution and the ability make "on the fly" changes during execution
- Design any number of the users own experimental workflows to suit their business needs using drag-and-drop functionality
- Execute: Simulate the experiment workflows before running to optimize execution, then start execution of the workflow with one-click operation
- Analyze: Track of experimental conditions in real-time to ensure optimal execution, and review experimental trends for in-depth analysis
- Report: Easily integrate with 3rd-party LIMS and ELN using the "Autsuite API" and powerful export functions for detailed reporting
- In-build capabilities such as drag-and-drop UI, task lists, connectivity, API, virtual testing, comprehensive data capture for import/export
- "Autosuite" provides various connections such as API, database, CSV import/export and execution of 3rd party program

The high-level software of Chemspeed offers specialized automated solutions for a variety of workflows – sample prep, synthesis, process research, formulation, application, testing. Multiple systems (incl. 3rd-party devices) are often required to fully automate and digitalize a partial or entire product development cycle. While "AutoSuite" software (Figure 8.2) provides an excellent control interface with the systems/modules, "Arksuite" provides another level of software control for orchestrating multiple instances (incl. 3rd-party devices) – global experiment design, experiment execution, scheduling, inventory management, data management, digital twin. Thus, it is an integrating software platform that serves as a central control centre. Whether instruments are located in the same or across multiple labs, programming one continuous workflow, that spans them all, provides the ultimate execution, schedule, inventory, and data management of the customers entire end-to-end (E2E) workflow. Integrating additional services such as laboratory information management systems (LIMS), web services, and other auxiliary tools further enhance the overall process. The inclusion of 3rd-party devices e.g. for analysis/testing completes the workflow, and now provides full E2E control and comprehensive data capture, the holistic step into the world of digitalization. Highly flexible design supports essentially any interface using IoT and Internet 4.0 concepts. The comprehensive experimental data capture includes universal execution time stamps. Central database services support ease of linking multiple systems, labs and even sites for uninhibited data share and collaboration (physical and/or cloud-based server solutions). The radical digital twin provides real-time top-level viewing and exceptional control of integrated systems and/or labs.

Labman's hardware all runs on their standardized "BaseApp" software, a fully modular app that allows for total freedom and even white labelling, whilst maintaining all of the inbuilt connections to hardware devices that they have developed over 40+ years of engineering. At this time, it is the best of both worlds, offering customizability, but without the hassle or the cost of starting software from scratch. The software team can also assist in developing mobile apps, data analysis tools and their solutions are fully compliant with many industry standards for data. The offering of seamless integration with LIMS of customers and many third-party databases, ensuring a unified platform for scientists to access and manage data across different sources. The integration allows for real-time data exchange from Labman's automated systems, reducing manual data

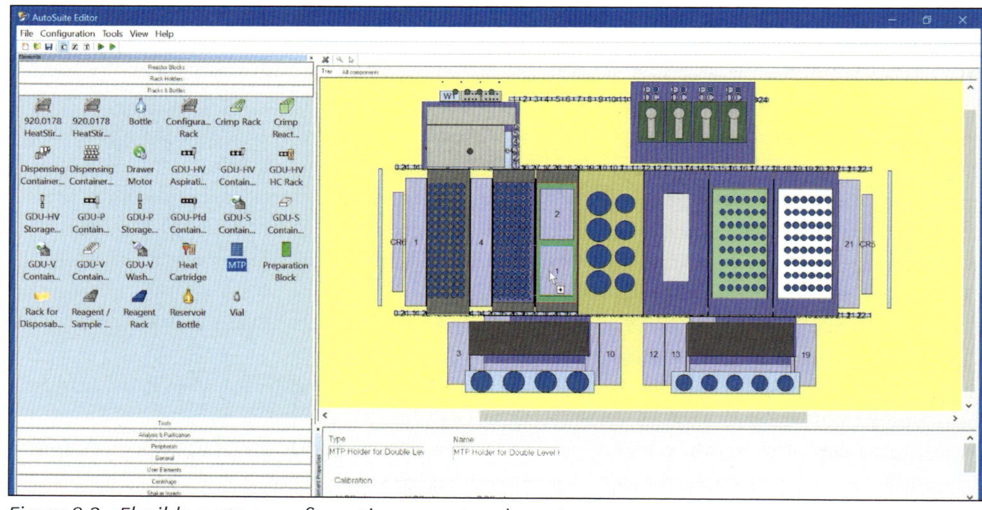

Figure 8.2: Flexible system configuration on system layout Source: Chemspeed Technologies

entry errors, and providing a comprehensive view of experiments, results, and sample information. Labman's software solutions can include features for digitalized experimentation, providing scientists with step-by-step protocols, automated data capture, and instrument integration. This ensures standardized and reproducible experiments, reducing variability and enhancing the reliability of experimental outcomes. Lab digitalization and electronic lab notebook (ELN) enabling multiple researchers to access and contribute to experiments with Labman's digitalization of laboratory processes. Integration with ELN's allows for scientists to capture, organize, and share experiment data in a digital format. The software and hardware are designed to scale with the evolving needs of laboratories. Hosting data locally or store it in the cloud, scalability and flexibility is offered to adapt to changing data management and security requirements.

Labman's "ControlApp" software connects almost any piece of laboratory hardware into a seamless user interface for Labman hardware, allowing for anything from fine robot controls to data discovery, all while being fully configurable to fit any brand or lab requirements. Creating fully custom software from scratch can be a long and expensive undertaking; Labman eliminates these problems with "ControlApp". Its fully customizable whilst retaining all functionality and connectivity that have been developed over the years. It is well known that many pieces of lab software are far from intuitive to use, and onboarding can be a serious headache. "ControlApp" is designed with UX principles in mind and is designed to be as seamless and understated as possible (Figure 8.3). It has been designed to be able to communicate with almost any LIMS systems available. For customers looking to integrate with third party software and hardware, a stateless REST API can be included for easy dynamic integration.

The roots of Füll Lab (former Bosch Automation) continue in using software PLC[1] ("Nexeed" automation from Bosch Connected Industries). They rely of the control on machine level, individual modules and systems via software PLC controllers (Bosch Rexroth or Beckhoff). These are running on industrial PCs which are integrated into separate electric cabinets (for larger systems) or directly on the individual modules or smaller machines. The software offers various standard APIs, which are implemented on a project-specific basis. Also, the "Nexeed" system offers a standard human machine interface (HMI) and refers to a dashboard that enables users to communicate with machines, computer programs or systems. HMI which is typically used for close-to-machine operation such as service and maintenance tasks. This HMI is also used for full machine control in close to production areas, such as the "BLS" spray application for quality control. When it comes to more complex workflows or if it is desired to combine results of different systems, Füll Lab Automation has an own software solution, the

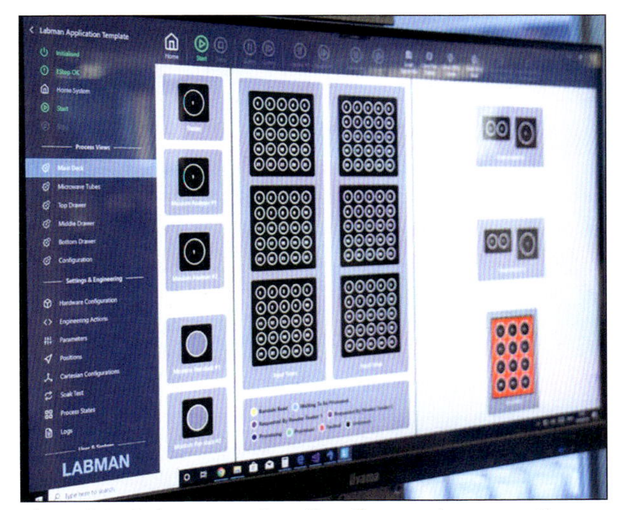

Figure 8.3: Labman user interface for sample preparation
Source: Labman Automation

1 Programmable logic controller

"Workflow Manager" (WFM). The WFM is a native Windows application which can be run on standard PCs or Servers. The WFM is the standard HMI for daily operation and therefore also allows to carry out basic maintenance and service tasks. It has an interface to the PLC controllers of Füll Lab´s machines but can as well be used to interface directly with metrology devices or even machines from other manufacturers. The WFM allows the operator to flexibly define and parametrize workflows or check the machine status. It offers the possibility to manage the storage of raw materials on the automated system as well as in the lab. Samples can be defined individually, and a combination of multiple samples is also possible. The latter is for example used when a base coat and a clear coat are both prepared on the system in the correct timely manner to coat a single panel. Data generated by the system (or external devices) are written directly to a database, typically an SQL database[2]. Connection to existing LIMS or MES systems is often implemented via a file interface but other communication protocols are also available, for example web-based data exchange. Besides exporting XML or CSV files, project specific data formats are possible. Furthermore, the database layout is open to customers and enables them to use direct retrieval of data through automated database queries. Finally, the WFM allows integration of or interfacing to AI and ML components and solutions, e.g. for parameter and formulation optimization. This enables the functionality to be extended and expanded with existing and future AI systems.

8.3 Artificial intelligence

Artificial intelligence (AI) is revolutionizing the development of paints and coatings by automating processes, analyzing data and promoting innovative solutions. From optimizing formulations to improving properties, AI offers numerous advantages for research and production. The paint and coatings industry plays a crucial role in many areas, as it has established a variety of products in very different markets. Although the industry has been around for a very long time, it is a dynamic sector characterized by continuous development and a wide range of applications. Technical innovations in the field of smart coatings are constantly being developed, for example antimicrobial coatings to improve hygiene, self-cleaning paints to minimize maintenance and special coatings for solar modules to increase efficiency. Two important trends that are currently shaping the paint and coatings industry are digitalization and sustainability. Manufacturers are increasingly turning to artificial intelligence and machine learning (ML) to improve various operational aspects, including research & development.

In the age of big data and digitalization, the application of AI in product development is becoming essential for manufacturers to remain competitive in their industry. Gone are the days when a limited group of professional's plans and execute Design of Experiments (DOE) until a set of target characteristics is achieved. R&D teams in many industries are transforming their workflows by using AI initiatives to intelligently extract information from their historical, experimental and product databases. AI models built on previous data can be used to reduce experimentation through simulation and achieve optimal properties much faster through numerical optimization. The possible applications of AI in the paint and coatings industry range from the development of formulations, colour analysis and colorimetry to materials research and quality control, sustainability and customer-oriented applications. Some common examples of the use of AI are:

- AI models analyze existing formulations and simulate new blends to achieve desired properties (e.g. colour intensity, shelf life)

2 Structured query language is a domain-specific language used to manage data, especially in a relational database management system

- AI suggests suitable raw materials based on databases and previous test results
- AI helps to find environmentally friendly and low-VOC (volatile organic compounds) solutions
- AI and machine learning algorithms calculate how certain pigment combinations affect color perception
- AI compares desired shades with existing databases and suggests adjustments
- Through simulation, AI-supported models can predict the mechanical, optical and chemical properties of coatings
- Discovery of new materials through AI analysis of large amounts of data to identify innovative binders, pigments or additives
- Image recognition systems with AI identify defects such as bubbles, cracks or unevenness in coatings in quality control
- For life cycle analysis (LCA), AI can evaluate the environmental impact of formulations and production processes
- When recycling paints, AI develops strategies for the reuse of residual paints and varnishes
- In the area of colour personalization, AI-based tools can enable customers to virtually test and customize colours
- AI can be used to analyze market data from social media and sales figures to predict new colour trends

AI brings many benefits to the paint and coatings industry. It enables predictive analysis of coating performance, increases production efficiency, optimizes formulations and improves customer satisfaction while reducing time and costs. For example, large language models (LLMs) can process large amounts of research data to identify trends or improve paint and coating formulations. They also help to improve customer loyalty through personalized recommendations, virtual consultations and integration into virtual and augmented reality systems. AI tools help to optimize ingredients in colour formulations and predict the impact of legislative changes, such as CO_2 emission limits.

AI is promising for the coating sector when viewed from different angles, such as resin design, coating performance, coating evaluation, prediction and defects. Some European paint manufacturers have already achieved considerable success in working with AI. The UK was one of the first countries to see the intersection between AI and contemporary colour design when HMG Paints launched its "Nature's Embrace collection" in 2023, which was shaped by the dual powers of ChatGPT as an AI source combined with the company's proprietary colour database – databases that are fundamental to most enterprise artificial intelligence applications. Recently, another application of artificial intelligence was revealed when AkzoNobel announced that it had embarked on a partnership with "coatingAI" with a view to developing its "Flightpath" technology for the powder coating sector [1]. The "Flightpath" software has succeeded by pairing artificial intelligence with comprehensive technical insights from AkzoNobel in order to achieve more flawless powder finishing, taking into account spray gun motion and varying conditions of application while at the same time harnessing recommended settings that have their roots in artificially intelligent design. It all sounds far too good to be true. What's the catch? There are challenges when using AI. One common problem is the existing data quality. AI requires extensive, high-quality data, which is often difficult to obtain in the coatings industry. This is the crucial point. The availability, not of high-quality data, but its scope and digital availability. This is precisely the starting point for laboratory automation, especially HTE, which must provide the extensive database. Without this, the results of AI will not be of high quality either. As a result, there must also be changes in research processes. Furthermore, the development and maintenance of AI systems also requires high investments and specialized experts. And last but not least, employee acceptance is crucial. They need to be trained to work with AI systems and build trust in the technology.

An example from Orontec (Figure 8.4) shows that you can never have enough intelligent input – especially when it comes to automating cognitive processes such as processing and editing information in a laboratory, for example. And this is exactly where "Q-Chain" process AI comes in. This AI is a software framework that can be used to automate laboratory processes in stages – regardless of whether you are working with an Orontec measuring system or other instruments and systems.

1. First, the process is mapped in a standardized format according to ISO 19510
2. Then all data sources are integrated, e.g. existing Excel tables, CSV files, ERP systems, but also data from measuring devices
3. Finally, the process is automated using a workflow engine

If desired, an AI module working in the background can ensure that the initiated automation process continues to learn and evolve. Orontec's approach is based on small modular systems (microsystems) as opposed to the monolithic approaches of large AI software providers. Only a small budget is required for this type of process optimization. According to Orontec, "Q-Chain" process AI delivers precise results very quickly: Projects are completed in less than a month. The system can be scaled as required and existing solutions can be easily integrated and replaced later if necessary.

In contrast to the "small" approach, large AI software providers such as Citrine are bringing out the big guns. For companies that can leverage AI using their platform, building a new formulation quickly, without compromising results, is standard operating procedure, using AI-guided experimentation, see Figure 8.5. The platform and its data storage make it easy to break down inter-organizational silos. Different people on different teams can all see what's been tested, what works, and what does not, leading to more effective research and experimentation. By making the stage-gate process more efficient, enables the path to market much more quickly. The customer can use their AI tools to determine which materials have a higher probability of working earlier in the discovery process. By making predictions for performance criteria earlier in the process, you can narrow your scope to a small subset of experiments, reducing time and money spent on poor outcomes.
How does such a platform work? First, The "Citrine DataManager" (CDM) enables to unify and leverage data across an enterprise. It is a chemistry-aware software toolkit to capture, enhance, analyze, visualize and communicate your data in a structured but flexible way. Data can be ingested quickly via CSV, Excel or directly using their API. CDM includes a flexible "Python" interface to

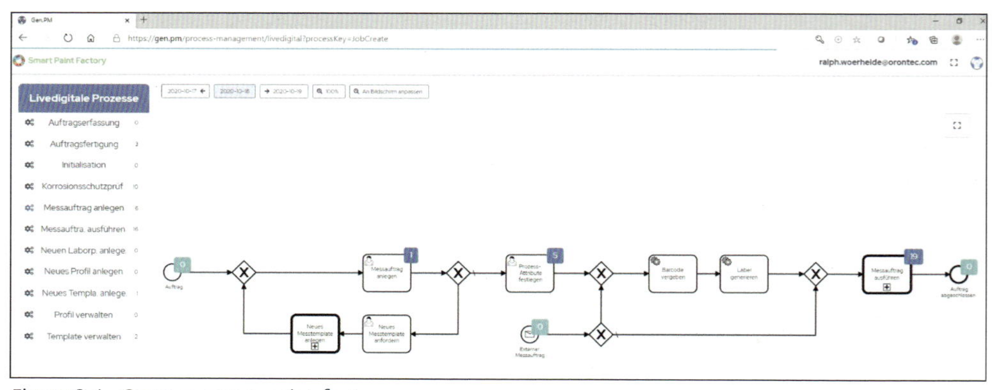

Figure 8.4: Orontec smart paint factory *Source: Orontec*

automate data ingestion. Second, the powerful AI system "Citrine Virtual Lab" (CVL) is conducting virtual experiments before the scientists go into the lab. CVL refers to a set of generative AI tools that allows to rapidly identify chemical formulations and materials that are likely to match a set of pre-defined specifications. The user needs to specify the customer needs, then the AI platform runs 1000s of virtual experiments and find the most promising candidates. By sequential learning workflows it powers the next generation of experiment design. Using iterative, proprietary calculations that factor in uncertainty, the user can efficiently and systematically see how each of the desired parameters might perform. The outcome of a recent case study with the international mineral filler manufacturer Dorfner in Germany, can be summarized as follows:

- Since adopting AI, sales across the company have grown by more than 30 %
- Formulations can be moved to new regions and maintain quality and consistency with local raw materials
- With AI predicting the most promising candidates, formulation development shrunk from 6 months to 1 month

8.4 Machine learning

Machine learning (ML) is a subset of artificial intelligence (AI) that enables systems to learn from data and improve their performance over time without being explicitly programmed. It involves creating algorithms that identify patterns, make predictions, or take actions based on data inputs. Common applications of machine learning can be found in healthcare, e.g. disease diagnosing using imaging and drug discovery, but also in finance, retail and e-commerce, autonomous systems such as self-drying cars, natural language processing, for example chatbot, and computer vision. Machine learning is increasingly transforming research and development (R&D) laboratories by automating processes, analyzing data, and driving innovation. In automated labs, ML plays a crucial role in enhancing productivity, accuracy, and decision-making. The process with the key elements for building an ML model is illustrated in Figure 8.6.

Figure 8.5: Integrating AI and real-time analysis in synthetic chemistry *Source: A Medium Corporation*

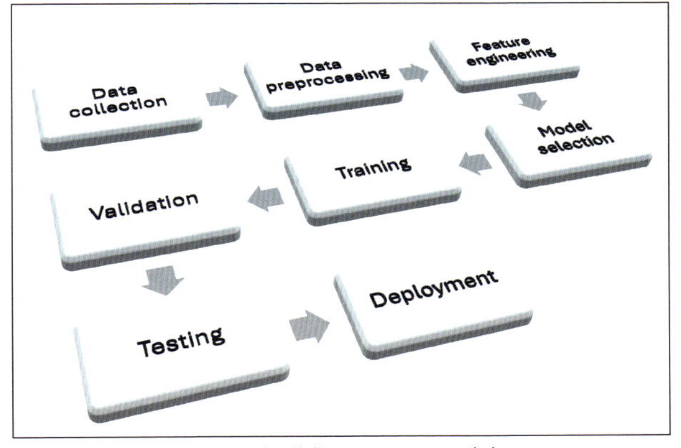

Figure 8.6: Key steps in building an ML model *Source: PERFECO Consulting Gysau*

Examples from industry and academia, but also cooperation, have been around for a long time. For example, the Berlin-based Joint Lab for Machine Learning (BASLEARN) was announced in 2019 [3]. It focuses on fundamental chemical issues, for example to develop practicable new mathematical models and algorithms from process or quantum chemistry. As an essential part of the cooperation, the BASF research group "Machine Learning and Artificial Intelligence" supported the research work for machine learning of the "Berlin Centre for Machine Learning" at the TU Berlin with a million-euro sum. The areas of application for machine learning range from biological systems to materials and drug research, laboratory automation and dynamic process systems. Specific examples of the joint research work include predicting the solubility of complex mixtures or colours as well as the aging processes of catalysts.

For engineers developing new materials or protective coatings, there are billions of different possibilities to sort through. Lab tests or even detailed computer simulations to determine their exact properties, such as toughness, can take hours, days, or more for each variation. A relatively new artificial intelligence-based approach developed at the Massachusetts Institute of Technology (MIT) could reduce that to a matter of milliseconds, making it practical to screen vast arrays of candidate materials [4]. The system, which the MIT researchers hope could be used to develop stronger protective coatings or structural materials – for example, to protect aircraft or spacecraft from impacts. The research was performed by MIT postdoc *Chi-Hua Yu*, Civil and Environmental Engineering Professor and Department Head *Markus J. Buehler*, and *Yu-Chuan Hsu* at the National Taiwan University. "Over the past 30 years or so there have been multiple approaches to model crack propagation in solids, but it remains a formidable and computationally expensive problem,"says *Pradeep Guduru*, a Professor of Engineering at Brown University, who was not involved in this work. "By shifting the computational expense to training a robust machine-learning algorithm, this new approach can potentially result in a quick and computationally inexpensive design tool, which is always desirable for practical applications." In the case having new material and a lot of images of the fracturing process they can feed that data into the machine-learning model as well. Whatever the input, simulated or experimental, the AI system essentially goes through the evolving process frame by frame, noting how each image differs from the one before in order to learn the underlying dynamics. By having more and more high-throughput experimental techniques with a lot of very fast, automated produced images these kind of data sources can immediately be fed into a machine-learning model. The future will be one where a lot more integration between experiment and simulation will happen, much more than in the past.

8.5 Literature

[1] KNOWLES, T.; "Recent Coatings Applications of AI; Coatings World, 7th Sept. 2024

[2] Citrine Informatics, "Dorfner partners with Citrine Informatics and Revolutionizes Paint Formulation Work through Artificial Intelligence", https://citrine.io/media-post/dorfner-partners-with-citrine-informatics-and-revolutionizes-paint-formulation-work-through-artificial-intelligence/, 2023

[3] FARBEUNDLACK, "BASF und TU Berlin kooperieren beim Thema Künstliche Intelligenz", www.farbeundlack.de/nachrichten/marktbranche/basf-und-tu-berlin-kooperieren-beim-thema-kuenstliche-intelligenz/, 10th Sept. 2023

[4] Paint & Coatings Industry, "Machine-Learning Research Could Help Develop Tougher Coatings", www.pcimag.com/articles/107543-researchers-use-machine-learning-to-help-develop-tougher-coatings, 12th June 2020

9 Economics of automation

Laboratory automation refers to the use of automated instruments, robotics, and software to perform routine and complex laboratory tasks with minimal human intervention. It significantly impacts cost, productivity, and overall efficiency in research, diagnostics, and industrial settings. The economics of laboratory automation involves understanding the costs, benefits, and long-term return on investment (ROI) associated with integrating automation technologies into laboratory workflows. The decision to introduce laboratory automation is based on strategic criteria. A thorough analysis of these criteria helps to select the right technologies and make optimum use of resources, see Figure 9.1. But what needs to be considered? First and foremost is often cost-effectiveness. Other criteria include increasing productivity, quality, standardization, data integration, flexibility, adaptability, employee integration and acceptance, availability of qualified employees, sustainability, environmental aspects, risk management, strategic goals and competitiveness and sure more. The decision for an investment therefore has far-reaching significance and must not only but must also be linked to economic efficiency [1].

9.1 Key economic drivers

In business, every investment, whether small or large, must be justifiably scrutinized, as it can have a variety of effects that may not be obvious at first glance. There are some key economic drivers to consider when investing in laboratory automation, regardless of the industry and application.

Increasing productivity is often mentioned first when it comes to high-throughput experimentation (HTE). Automated systems perform repetitive tasks faster and more consistently, thus enabling higher sample throughput. Higher throughput means achieving more experiments and results in the same amount of time. More results in the same period of time enable developments to be accelerated or, in other words, the development time and thus the time to market to be shortened. Every economist knows what this means. A shorter development time leads to faster achievement of sales targets. In addition, as is well known, the frontrunner has a larger share of the overall market in the long term. The pleasant side effect of increasing productivity is the creation of free capacity for laboratory staff. Instead of mindless routine work, they can focus on more challenging activities and perhaps even greater added value. This

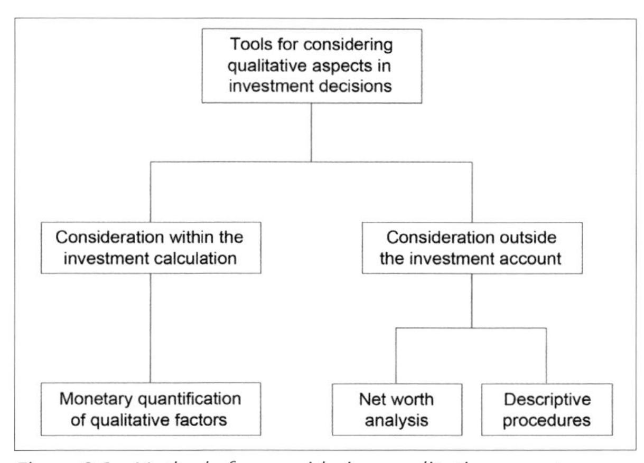

Figure 9.1: Methods for considering qualitative aspects

increases employees' self-esteem and motivation. It also boosts creativity and innovation and, in general, the attractiveness of the individual employee's job and the company's reputation for future recruitment.

Cost efficiency is another driver for laboratory automation. The unattended operation of a laboratory automation system, apart from daily maintenance tasks such as replenishing the HTE with consumables and raw materials, reduces labour costs by minimizing manual intervention. Reduced daytime labour extends to unattended overnight and weekend use, depending on the configuration of the laboratory automation. Laboratory automation is accompanied by a reduction in the size of experiments because automation leads to an increase in precision. The batch size can be reduced to the minimum quantity required for the tests to be carried out. Preparation sizes of 100 to 150 g are often sufficient for HTE. By significantly reducing the batch size, the consumption of raw materials is also significantly reduced. After completion of the workflows, only a minimal amount remains, which represents the safety margin for carrying out all necessary experiments. In summary, raw material consumption and waste per experiment is significantly reduced, resulting in cost savings.

As mentioned above, **improved accuracy and precision** is another driver for key economics. Accuracy increases significantly with automation, as many gravimetric dosing systems work with a balance resolution of between 0.1 and 10 mg. When weighed by humans, apart from in analytics with higher resolution scales, the most precise manual weighing in the paint laboratory is in the range of 10 to 100 mg. This means that the precision in automated operation is 10 to 100 times higher, which increases the reliability of the results and reduces the number of repeat tests due to uncertain, not always explainable results in manual operation.

Scalability contributes as well as a driver for key economics. Additional order volumes can generally be absorbed by a high-throughput experimental system, depending of course on the system's degree of utilization. A capacity utilization rate of 100 % is rather the exception, but not impossible. In addition, urgent orders can also be prioritized in an HTE, which leads to faster processing. These advantages are possible without additional human resources in contrast to manual laboratory operation.

It is rather unusual in paint and coatings laboratories, but in other industries, **regulatory compliance** would be an additional driver for automated laboratory operations. The complete documentation, standardization and traceability provided by automated systems helps laboratories meet stringent regulatory requirements efficiently.

9.2 Costs associated with laboratory automation

CAPEX (capital expenditures) and OPEX (operational expenditures) are key terms in financial and operational management, especially when planning and analyzing projects, investments or technologies. They describe the two main types of expenditure incurred by a company. In the context of investments in laboratory automation, which is usually costly, it does no harm for scientists to take a brief excursion into the world of finance at this point. This may make it easier for them to justify their CAPEX in their next expenditure application.

The definition of CAPEX includes all long-term investments made for the acquisition, improvement or construction of assets. These investments are necessary to secure or expand a company's ability to operate. In the following is an example of the costs of implementing an automation system with the associated features, which are generally applicable.

Table 9.1: Differences between CAPEX and OPEX

Criteria	**CAPEX**	**OPEX**
Type of costs	One-off or infrequent expenses	Recurring, ongoing expenses
Period	Long-term benefit	Short-term, within the period
Accounting treatment	Depreciation over useful life	Direct charge to income statement
Flexibility	Low, long-term commitment	High, adjustment possible
Objective	Build-up of assets	Maintenance of operations

- One-off costs: CAPEX is usually incurred as a major, one-off expense
- Depreciation: These investments are depreciated over their useful life and are not immediately charged to the income statement
- Long-term benefits: The investments should create added value over several years

What are the investment challenges?
- High capital requirement
- Long-term planning and financing required
- Risks associated with large investments if market conditions change

The definition of OPEX includes all ongoing expenses that are necessary to maintain the company's operations. They are recurring and often directly linked to day-to-day business. Some examples are listed below.
- Wages and salaries
- Costs for utilities such as electricity, water and gas
- Maintenance and repair of machinery
- License fees for software
- Costs for consumables such as raw materials, reagents or office supplies

What are the associated features with OPEX?
- Recurring costs: These costs are incurred on a regular basis, often monthly or annually
- Direct profit burden: OPEX is recognized directly in the income statement
- Flexibility: In contrast to CAPEX, OPEX is often easier to adapt to changing conditions, see Figure 9.1

The challenges for OPEX lie in controlling costs in the face of rising consumption and the need to increase efficiency in order to remain competitive.

A CAPEX-focused strategy is suitable for (large) companies with stable financial resources, a focus on long-term assets and sustainable benefits. An example is the establishment of a new laboratory consisting of standalone automation solutions or an HTE. This contrasts with the OPEX-focused strategy, see Table 9.2. This is preferred by companies that require flexibility and lower initial investment. Very often, these tend to be small and medium-sized enterprises (SMEs). Either smaller automation solutions which can be modular expended or outsourcing or "pay-per-use" models (e.g. rental solutions for devices) are therefore an option for them. In the field of laboratory automation, there are already financial service providers that offer leasing models.
What other considerations help in the decision-making process? A detailed calculation of the return on investment (ROI) helps to evaluate the cost-effectiveness of automation. It is also advisable to carry out a cost-benefit analysis. This compares long-term investments with ongoing costs in order to select the best strategy. Lifecycle management looks at the total costs over the lifetime

Table 9.2: CAPEX and OPEX in laboratory automation

CAPEX	OPEX
Acquisition of automated equipment (e.g. standalone automation, HTE)	Maintenance and calibration of equipment
Installation of laboratory information systems (LIMS)	Consumables (reagents, syringes, tips, cartridges)
Building a new facility, because of existing space limitations	Energy consumption of automated systems
Setting up an infrastructure (e.g. special laboratory ventilation or energy supply)	Training and further education of employees
	Software updates and license fees

of a project or device and provides an expected financial summary. Which financing models can be considered? Leasing or rental options can turn high CAPEX expenditure into OPEX and allow SMEs to balance their expenditure more evenly.

9.3 Return on investment (ROI)

ROI is a critical metric used to evaluate the financial benefits of laboratory automation relative to its costs. It helps laboratories and organizations decide whether investing in automation technology is justified and profitable over the long term. The ROI for laboratory automation depends on factors such as workload volume, complexity of tasks, and long-term cost savings. How is the ROI defined and can be calculated, see Equation 9.1.

$$\text{ROI (\%)} = (((\text{Net benefit (Total gains} - \text{Total Costs)}) / \text{Total costs}) \times 100 \quad \text{(Equation 9.1)}$$

Definitions:
- Net benefit: Savings and revenue improvements minus operational expenses.
- Total costs: Includes both CAPEX (capital expenditures) and OPEX (operational expenditures) related to automation.

The components of ROI for laboratory automation have been already discussed in Chapter 9.2, Table 9.2. The benefits (return side) of direct and indirect gains are listed below.
- **Direct financial gains:**
  - Increased throughput (more samples processed in less time)
  - Reduced labour costs by automating repetitive tasks
  - Savings on reagents and materials through precise dispensing
  - Faster time-to-market for R&D projects
- **Indirect gains:**
  - Improved data quality and reduced errors
  - Higher customer satisfaction through faster turnaround times
  - Regulatory compliance with less manual documentation effort
  - Mitigation of risks associated with human error

The success criteria of the ROI depend as well on the applied time period which may be different from industry to industry and the type of investment. Three common types are described as follows.
- **Short-term** (1–2 years)
  - ROI may appear low due to high initial investment

  - Benefits are primarily seen in reduced errors and improved efficiency
- **Mid-term** (3–5 years)
  - Operational savings and throughput improvements become evident
  - Enhanced scalability allows revenue generation from increased capacity
- **Long-term** (>5 years)
  - Savings from lower labour costs and reduced downtime dominate
  - Equipment depreciation reduces the perceived cost of ownership

ROI example calculation (scenario):
A R&D coatings lab is considering automating its rheology measurement workflow.
- Investment (CAPEX): EUR 500,000 for automated equipment
- Annual operational costs (OPEX): EUR 50,000 for maintenance, consumables, and staff training

Annual benefits:
- EUR 180,000 saved in labour costs
- EUR 30,000 saved in reduced reagent waste
- EUR 100,000 in additional revenue from higher throughput

Calculation:
- Total costs (5 years) = EUR 500,000 + (EUR 50,000 × 5) = EUR 750,000
- Total benefits (5 years) = (EUR 180,000 + EUR 30,000 + EUR 100,000) × 5 = EUR 1,550,000
- Net benefit = EUR 1,550,000 – EUR 750,000 = EUR 800,000

$$ROI = (800{,}000 / 750{,}000) \times 100 = 107\ \% \qquad \text{(Equation 9.2)}$$

The ROI of 107 % (Equation 9.2) indicates the investment more than doubles the returns over five years.

Calculating ROI for laboratory automation involves a balance of upfront costs and long-term savings and gains. With careful planning, most laboratories can achieve positive ROI within a few years, driven by enhanced efficiency, reduced errors, and increased throughput. Tailoring automation investments to specific laboratory needs ensures maximum financial and operational benefits.

9.4 Challenges

While laboratory automation offers significant benefits, implementing and maintaining these systems can present several challenges. Understanding these challenges is essential for successful integration and operation, especially if it is the first investment into laboratory automation and thus starting the digital transformation of companies.

A CAPEX for a laboratory automation system requires **significant upfront investment** in equipment, software, and infrastructure. Smaller laboratories may struggle to justify these costs without guaranteed throughput or cost savings. Expenses for installation, training, and integration with existing systems are often underestimated or simply forgotten.

The integration of a new laboratory automation system may cause **compatibility** issues as they may not seamlessly integrate with legacy equipment or software. Incompatibility can lead to inefficiencies or the need for costly upgrades. From the data management perspective, the integration into laboratory information management systems (LIMS) or enterprise resource planning (ERP) tools can be complex. The provision of proper application programming interfaces (API) to connect the automation workflow with existing software applications is mandatory, but quite often fall under the radar. Ensuring consistent data transfer and real-time updates is also a common hurdle.

Some automation systems are designed for specific tasks with fixed protocols and **lack adaptability** for new or varied processes. This can limit their utility in research or multi-purpose laboratories where tasks frequently change. Adapting automation for unique laboratory workflows may require expensive modifications or software development and considered under customization challenges.

As the size of laboratory automation increases, so does the degree of **technical complexity**. This has a direct impact on maintenance and downtime. Automated systems require regular maintenance and calibration to function optimally. Unexpected technical issues can cause downtime, disrupting workflows and increasing costs. In addition, advanced technical knowledge is often needed to operate and troubleshoot automation systems. A lack of skilled personnel can delay implementation and reduce system efficiency.

One of the often-notified challenges is the **resistance to change**. Staff may fear job displacement or feel intimidated by new technologies. This can lead to resistance, reducing the effectiveness of automation initiatives. One of the solutions is to involve the employees intended to operate in the future the laboratory automation system as early as possible. The involved staff need to have a positive attitude or even better a passion for automation. Significant time and resources must be invested in training staff to use automated systems effectively.

The use of laboratory automation aims to increase sample throughput and also reveals **scalability challenges**. Laboratories with fluctuating workloads may find it difficult to scale automated systems up or down efficiently. Over-automation can lead to underutilized equipment and wasted resources. If the system is not used to its full capacity, the expected benefits may not materialize, reducing cost-effectiveness. Besides the workload challenges, there might be also space constraints. Automated systems, especially large ones, often require significant physical space, which may not be available in smaller labs.

Due to the increasing threat of cyber-attacks, internal IT requirements for **data security and privacy** are rising. However, automated systems often rely on interconnected networks for storing data on company servers or clouds, making them vulnerable to cyberattacks. Protecting sensitive data (e.g., confidential information, intellectual property) requires robust cybersecurity measures. Automation generates large volumes of data, necessitating efficient storage, management, and analysis solutions.

Challenge and risk in common is the **dependance on vendors**. Many automated systems are proprietary, leading to dependence on a specific vendor for upgrades, repairs, and consumables. This can increase long-term costs and limit flexibility. In case of inadequate technical support from vendors, this can delay troubleshooting and impact laboratory operations with a direct impact on uptime and speed of innovation.

As soon as the decision for budget and CAPEX is received from management, the automation system shall be ready as soon as possible, but often it is rather a **long implementation time**. Depending on project complexity, the implementing of automation requires detailed planning, including workflow analysis, equipment selection, project execution with define & design phase, engineering, procurement, assembly, commissioning, testing for factory acceptance test (FAT), delivery and site acceptance test (SAT), and staff training. This process can be time-consuming and delay the realization of benefits. And, adjusting to automated workflows takes time, potentially slowing initial productivity in relation to the learning curve.

Environmental and sustainability concerns may affect the company's overall strategy, e.g. in terms of energy consumption and waste management. Automation systems often have high energy demands, increasing operational costs and environmental impact. Automated systems may generate more plastic waste due to an increased consumption of consumables, particularly in high-throughput environments.

Despite all the challenges, laboratory automation and in particular HTE have their fascination and thus existing strategies to overcome these challenges.

- Comprehensive planning: Conduct a detailed cost-benefit and workflow analysis before implementation. Prioritize automation for high-volume, repetitive, and error-prone tasks.
- Stakeholder engagement: Involve staff in decision-making to address concerns and ensure smoother transitions. Provide thorough training to build confidence in using automated systems.
- Vendor selection: Choose reliable vendors with proven support and scalable solutions. Negotiate service-level agreements (SLAs) for maintenance and updates.
- Scalability and flexibility: Choose modular systems that allow phased implementation and future expansion.
- Data management solutions: Invest in robust LIMS and cybersecurity to handle data integration and security effectively.

The economics of laboratory automation highlight the balance between upfront investments and long-term savings. By increasing productivity, enhancing data quality, and reducing operational costs, automation offers significant advantages. While the initial investment can be prohibitive for some laboratories, careful planning and strategic integration often result in a strong ROI, particularly for high-throughput or complex workflows.

9.5 Net worth analysis

The net worth analysis (NWA) is a structured decision-making process that is used to evaluate and compare qualitative and quantitative criteria. It is particularly suitable for complex decisions in which several alternatives have to be evaluated on the basis of different evaluation criteria.

The data that is difficult or impossible to record in monetary terms is evaluated separately, independently of the investment calculation. All criteria relevant to the decision are listed and weighted according to their importance. The alternatives under discussion are then assessed on the basis of these criteria. For example, a scale of 1 to 6 is defined for each characteristic. This gives the score that corresponds to the unweighted benefit of the variant in relation to the respective criterion. The product of weighting and evaluation products corresponds to the weighted benefit of the respective variant with regard to a criterion. The sum of the weighted individual benefits gives the total benefit of the investment.

In the example shown here, two systems are used which differ not in terms of the provided quote, but in terms of qualitative arguments such as user-friendliness, flexibility, ecology and supplier reliability. The following value scale is used to determine the unweighted benefit of the various options, see Table 9.3.

The company attaches particular importance to flexibility and supplier reliability. Based on the resulting weighting of the individual criteria and the unweighted benefits of the options, the project manager determined the following ranking of the variants, see Table 9.4.
The decision as to which investment is to be preferred is made on the basis of the differences between the target values of the investment calculation and the difference in utility value.

When selecting the criteria, care must be taken to ensure that the characteristics are independent of each other and that the characteristics are not recorded more than once. In addition, only

Table 9.3: Value scale of unweighted benefits

Criteria	1	2	3	4	5	6
User friendlyness	Extremely poor	Very poor	Poor	Good	Very good	Excellent
Flexibility	Very small	Small	Relatively small	Relatively large	Large	very large
Ecology	Extremely poor	Very poor	Poor	Good	Very good	Excellent
Supplier evaluation	Extremely poor	Very poor	Poor	Good	Very good	Excellent

Table 9.4: Weighted ranking of supplier A and B

Criteria	Weight	Supplier A Evaluation	Supplier A Benefit	Supplier B Evaluation	Supplier B Benefit
User friendlyness	20	2	40	5	100
Flexibility	40	6	200	3	120
Ecology	10	3	30	4	40
Supplier evaluation	30	3	90	4	120
Total points	**100**	**360**		**380**	
Ranking		2.		1.	

those options should be subjected to a value-in-use analysis that fulfill all framework conditions and mandatory criteria. The framework conditions may also include minimum net present value (NPV) or internal rate of return (IRR) values. As both the weighting of the criteria and the classification of the variants in the value scale are subject to subjective assessment, a subsequent sensitivity analysis of the critical factors is indicated in some cases [1].

9.6 Literature

[1] GUREVITCH, D.; "Economic Justification of Laboratory Automation"; JALA: Journal of the Association for Laboratory Automation, p. 33ff, volume 9, issue 1, February 2004

[2] SEILER, A.; "Financial Management – BWL in der Praxis II"; p. 439ff, Edition 4, Orell Füssli Verlag, Zurich, 2007

[3] GOETZE, U.; BLOECH, J.; "Investitionsrechnung"; p. 173ff, Springer-Verlag, Berlin, 2003

10 Outlook

The global lab automation market size is estimated to grow from USD 5.5 billion in 2023 to USD 16 billion by 2035, representing a CAGR of 9.3 % during the forecast period 2023 to 2035. Increasing shortage of (skilled) manpower and high demand for testing and characterization plus to cope with increased regulatory pressure are major factors driving future market growth. The drivers of laboratory automation also include the ever-increasing R&D spending by blue chip companies to secure and even grow their market position. The screening of raw materials and development of innovations such as smart coatings are time-consuming development cycles that involve high costs and a high potential for risk, especially from human error. Automation of individual processes in this area leads to a drastic reduction in time and enables a significant reduction in human errors.

The U.S. has the largest overall market share, due to increasing spending and high prioritization of R&D opportunities in both the public and private sectors. In the European market, the UK and Germany lead the way of laboratory automation. In Asia, China is by far the largest market for laboratory automation. Japan is traditionally one of the leading countries in the field of laboratory automation. Very high annual growth rates of 8.6 % are expected for South Korea. As the trend toward connectivity and automation in R&D accelerates, connected labs will increasingly rely on intelligent networks, more sophisticated AI capabilities, and real-time collaboration tools. The goal is to create labs that not only perform R&D efficiently but are also adaptable, self-optimizing, and capable of scaling with minimal manual intervention. These advancements will likely make connected automated labs the standard for high-performance, data-centric research environments.

10.1 Future trends

The pioneers made the first investments in the noughties. Among them, for example, are the companies AkzoNobel in Slough/UK, BASF in Ludwigshafen/DE and Mankiewicz in Hamburg/DE. Many followers then joined them in the years from 2010 to 2020. Many other companies have recognized the potential and also put out their feelers for laboratory automation. However, wishful thinking often conflicts with the available budget, which is why an investment was postponed for the time being. It may also have been that the scope of the investment was reduced, and the focus was placed on the real bottlenecks. Standalone solutions are ideally suited for this, as they can be expanded with additional functions at a later date and, under certain circumstances, interconnected. No matter how big or small the budget may be, it is important to focus on the real bottlenecks and the large number of recurring routine tasks, both today and in a few years' time.

It is also important to take future trends into account as far as possible. Generally speaking, the main future trends that can currently be identified are:

4. **AI and ML:** Enhancing automation by integrating predictive analytics for experimental optimization
5. **Miniaturization:** Compact automated systems make advanced automation accessible to smaller laboratories
6. **Remote monitoring:** Cloud-based systems allow remote control and data analysis
7. **Cost reduction:** Technological advances and economies of scale will make automation more affordable

The general trends can be further subdivided into topics such as digital solutions, software solutions for run execution and workflows, AI and ML.

- Obviously, **edge computing** is rapidly evolving as a critical enabler for real-time processing, reduced latency, and improved efficiency in modern digital ecosystems. By processing data closer to its source, edge computing complements cloud infrastructure, enhancing performance in industries like IoT, AI, and 5G (Fifth Generation).
- **5G** is the latest generation of cellular network technology, designed to significantly enhance wireless communication. It offers unprecedented speeds, low latency and the ability to connect a large number of devices simultaneously. It is more reliable and faster. This transformative technology is reshaping industries, enabling innovations like smart cities, autonomous vehicles, and immersive augmented reality.
- **Hyper automation** is a business-driven approach that integrates advanced technologies, such as AI, ML, and robotic process automation (RPA), to automate processes across organizations more comprehensively, e.g. end-to-end automation. It aims to go beyond traditional automation by creating systems that not only automate tasks but also analyze, adapt, and improve continuously. This concept is a cornerstone of digital transformation and is recognized as a key trend by *Gartner* in recent years.
- **Augmented reality** (AR) and **virtual reality** (VR) are immersive technologies that are reshaping industries by enhancing how people interact with digital content. They enhance training, design, and customer experiences. While AR overlays digital elements on the real world, VR creates entirely virtual environments.

Artificial intelligence software continues to evolve rapidly, driving innovation across industries. As technology matures, several trends are emerging that will shape the future of AI software development and application. The number of developments of tailored AI solutions for specific industries will increase. There are certain industries that will benefit more than the paints and coatings industry, e.g. healthcare, manufacturing, finance and retail. Furthermore, the degree of transparency of AI decision plus the focus on responsible AI development and use is considered as trend required from the user community. A further trend is the expanded use of AI to automate complex workflows as used in automated R&D laboratories.

New trends typically generate market growth. Many markets would be happy with a high single-digit annual growth rate. In contrast, the annual growth rate for top AI software is almost dizzying. In the fiscal year 2022, the global valuation of the artificial intelligence sphere touched a monumental sum of USD 136.55 billion. Projections indicate an exponential escalation, with a Compound Annual Growth Rate (CAGR) of 37.3 % anticipated between the years 2023 and 2030[1].

Machine learning-driven lab automation is at the forefront of revolutionizing research and development in scientific laboratories. By enabling smarter, more adaptive systems, ML enhances automation by optimizing workflows, accelerating experiments, and uncovering insights from vast datasets. The most important future trends that will shape this area are described in the following.

- **Autonomous experimentation** is a trend, which has already a proven history, for fully automated systems that design, execute, and analyze experiments with minimal human intervention. Key features of this trend are ML models predicting optimal experimental conditions, real-time adjustments based on intermediate results and integration of robotics for precise sample handling and processing.
- The topic **preventive maintenance of lab equipment** will benefit by ML algorithms forecasting equipment failures and optimizing maintenance schedules. Examples are the

1 Source: Markovate.com

monitoring of sensor data (e.g., temperature, vibration) and anomaly detection to predict breakdowns before they occur. This minimizes downtime and prevents costly disruptions and thus extends the life span of critical lab instruments.

- ML-driven tools optimizing complex, multi-step laboratory processes. It allows identifying in **AI-powered workflows** bottlenecks and inefficiencies. The derived recommendations lead to sequence optimizations for faster execution. As a result, throughput will improve in high-demand laboratories, while resource waste and operating costs will decrease.
- ML systems automating the extraction of insights from large, multidimensional datasets for **advanced data analysis and visualization**. This allows pattern recognition in experimental results, advanced visualization tools for intuitive understanding and the integration with lab notebooks and data management systems. The impact is an accelerated hypothesis validation and supported reproducibility and transparency in research.
- The trend for **adaptive, dynamic quality control** and monitoring using ML models during experiments or manufacturing. This enables real-time analysis of process parameters and early detection of deviations from desired outcomes. Based on these features ML ensures consistent quality in lab outputs and reduces waste by detecting errors earlier in workflows.
- **Digital twins** create virtual replicas of laboratory systems, enhanced by ML, featuring real-time simulation of experiments and workflows as well as ML-driven predictions of outcomes based on virtual trial runs. This has a significant impact on the reduced need for physical trials, saving time and resources plus enables rapid testing of "what-if" scenarios.
- **Democratization of automation** with low-code platforms empowering labs with ML automation through low-code/no-code interfaces. Examples are drag-and-drop workflows for designing automated experiments, which are already implemented in existing HTE's, or pre-trained ML models integrated into lab software. This makes advanced automation accessible to non-programmers and speeds up adoption of ML-driven tools in smaller labs.
- **Cloud-connected lab platforms** using ML for collaborative analysis and remote automation. Shared datasets, experimental designs and real-time updates on ongoing experiments can be shared across global teams. The impacts can be considered extraordinary as they accelerate global research efforts and reduce the need for physical presence in labs.
- Enhancing human expertise with AI-driven recommendations and insights for the **human-machine collaboration**. ML suggesting next best actions based on experimental progress plus seamless human override and input for creative decision-making, see Figure 10.1. The balance between automation and human intuition is maintained and encourages adoption by augmenting rather than replacing scientists.
- **Augmented reality** interfaces integrated with ML to enhance lab operations. This features the real-time visualization of experiment data via AR glasses and provides step-by-step augmented guides for using complex instruments. This disruptive technology has been already presented at the European Coatings Conference in 2023 [1]. This allows reduced training time for new personnel, overcomes lack of qualified lab personnel and improves accuracy in multi-step procedures.

Technological trends on the hardware side are more likely to be seen in the area of logistics and the integration of more and more measuring instruments in fully automated laboratories. The area of high-precision dispensing of liquids and solids is already established on the market and is state of the art. As a result, few innovations are to be expected here. In contrast, the number of measuring instruments is almost infinite. Common measuring instruments for determining optical properties such as colour values and gloss have been an integral part of the automated development of paints and coatings for a good two decades, regardless of whether they are small handheld or larger tabletop devices. In automated R&D systems for raw material manufacturers or for quality control, analytical methods such as chromatography, spectrophotometry and spectroscopy or particle size measurement are absolutely standard. Recent automation examples show the integration of complex application methods such as roll-to-roll and spray with very high throughputs. There are also examples of mechanical-dynamic measurement methods such as nanoindentation, tensile and tensile shear tests, dynamic mechanical analysis and solid rheology. Other methods for the characterization and analysis of paint films and raw materials include scanning electron microscopy, X-ray diffraction and titrations will continue to grow as integral parts. The area of powder coatings and other special coatings and niche applications appears to be problematic because no automation solution has yet been implemented, but perhaps only because of the comparatively lower scalability in the market.

In the field of logistics in automation solutions, autonomous mobile robots (AMR's) will certainly increase in the future. Until now, providers have connected intralogistics between automated functionalities using sliders, robotic transfers static or mobile on a linear axis or with shuttles on a rail system. In some cases, manual logistics solutions are also used, sometimes with technical aids such as trolleys, partly because they are inexpensive. What they all have in common is the connection of functional modules or standalone automation systems. Self-driving robots and autonomous floor vehicles, equipped with sensor technology to prevent collisions, can now be found in many everyday situations, whether in production or warehouse systems, in restaurants or hotels. Their number, especially in public spaces, is still small, which is why there are often astonished looks. Approaches in barrier-free chemical laboratories are not new and lend themselves as a connection to standalone systems. Of course, there are already approaches for the use of quadcopters. However, these have so far mostly failed for safety reasons but offer good opportunities for logistics between laboratories that are scattered over a larger factory area or in chemical parks.

Figure 10.1: Robot human interaction in the lab of the future *Source: stockcake.com*

10.2 Supply situation

The three established and leading full-service providers and system integrators of laboratory automation, Chemspeed Technologies, Füll Lab Automation and Labman Automation, will seek to consolidate and further expand their market presence. Mature technologies, decades of experience and ongoing, continuous development will not change their market leadership in the near future. In addition to hardware, software plays an enormously important role and will become even more significant in the future. The challenges lie in particular in linking the hardware and run execution and workflow management software already in use with ELN, LIMS, ERP and, in particular, AI & ML. The influence of software will increase even more and require the use of more resources. To focus on this specialist sector, Bosch Packaging Technology with Bosch Lab Automation became Syntegon in 2020. Just one year later, the Lab Automation division was spun off into Füll Lab Automation in Ostfildern near Stuttgart, DE. Chemspeed began expanding the company several years ago by moving to much larger premises. In 2024, Chemspeed Technologies was acquired by the well-known instrument manufacturer Bruker and has since operated as an independent vendor agnostic automation division within the Bruker BioSpin business unit. The last of the group, Labman has recently completed a major expansion of its site in North Yorkshire, UK. Recent developments show, besides their capabilities in dosing, application and testing a strong focus on the user experience. This means that the companies are fit for the future.

There are two different types of providers of standalone systems for the automation of individual functions and possible additional options. On the one hand, instrument manufacturers supplement their existing portfolio with automated sample preparation prior to measurement with their instruments. In the past, the only type of automation was, for example, a sample carousel for feeding samples into the measuring cell. This includes the classic analytical measuring devices for chromatography, spectroscopy and particle measurement. This relieves the burden on laboratory staff, who previously had to perform the usually time-consuming sample preparation manually. On the other hand, the full-service providers naturally do not want to miss out on this business. In the field of viscometry and rheology in particular, there are currently system solutions from Anton Paar on the instrument manufacturer side, as well as solutions from Chemspeed, Füll Lab and Labman. The market for instrument automation will grow significantly over the next few years, if only due to the problem of recruiting qualified specialists. However, the biggest driver is the digital transformation and the utilization of data for AI and ML and among globally active teams. From a strategic perspective, this has certainly also prompted Bruker to acquire the global automation manufacturer Chemspeed. This also fits in with its investment in software solutions, which led to the creation of their software division called "SciY". Acquisitions among paint and coatings manufacturers have been commonplace for decades, but a new chapter is being opened in the field of automation. It remains to be seen whether this will continue in the coming years. Strategic partnerships or acquisitions between large laboratory instrument manufacturers and software companies in the field of AI & ML cannot be ruled out either.

In addition to many examples of automated R&D applications, the market for quality control (QC) is still weak, see Figure 10.2. But it is precisely here that automation offers enormous potential to come in the future. The standardization and reproducibility that can be shared in real time in internationally operating teams should be emphasized. This also makes it possible to compare products manufactured at different locations within a group of companies [2]. Another important point is the harmonization of QC operations with the production cycle, which is often already automated in 24-hour operation. Automated, synchronized 24-hour QC operation prevents laboratory backlogs in the morning, especially on Monday mornings after the weekend. This speeds up pass/fail decisions and accelerates the further process in production and the subsequent logistics chain all the way to the customer.

The number of Asian suppliers, especially from China, is still very modest at the moment, but it will happen. The first steps have been taken in the field of automation for the characterization of paints and coatings with standalone systems, but the systems are still somewhat rudimentary, see Figure 10.2 to 10.3. Robustness, precision and user experience have some catching up to do. The breadth of the portfolio is also still very modest, but as in other markets, it can be assumed that China will learn quickly. However, dosing, dispersion and application technology will be decisive for the paints and coatings industry. This is also the basic prerequisite for becoming a full-service provider for end-to-end laboratory automation. And finally, paints and coatings lab automation still require a high degree of customization, which makes it difficult for scalability which is the business model in China.

10.3 Lab of the future

The "Lab of the Future" (LoTF) is a concept driven by technological innovation, digital transformation, and sustainability. It envisions laboratories as intelligent, interconnected ecosystems designed to maximize efficiency, enable groundbreaking discoveries, and adapt to evolving research and operational needs. The Lab of the Future contains several key characteristics driving innovation to the next level, see Figure 10.4.

- **Fully connected ecosystem** with seamless integration of devices, systems, and data via the internet of things (IoT) will have centralized management of workflows and data through cloud-based platforms. Connectivity between global teams for collaborative research will happen in real time with automation at all levels.
- The application of **smart robotics** with automated sample preparation, characterization, testing, and complex analyses will accelerate innovation. End-to-end automation of experiments,

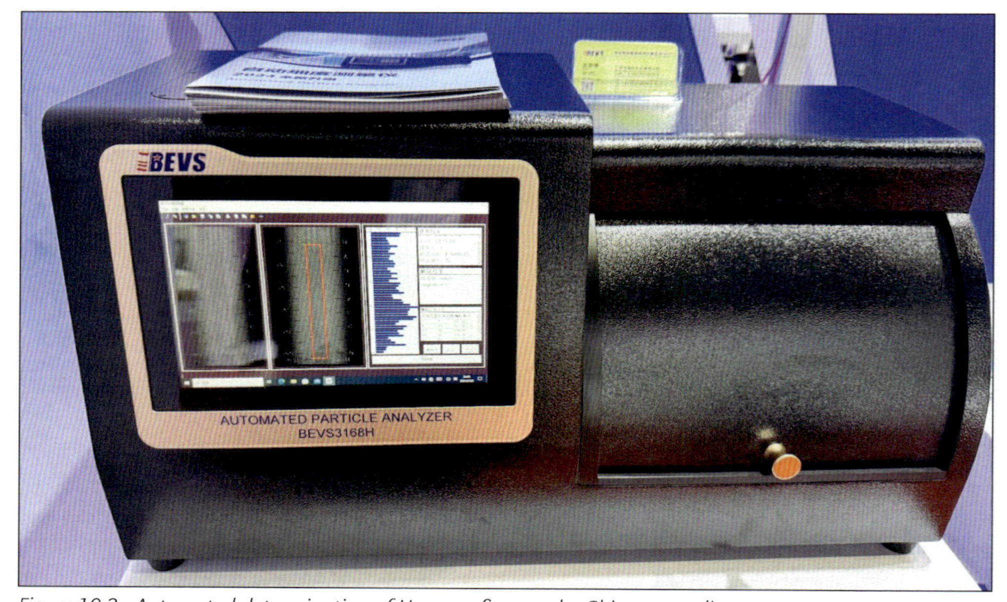

Figure 10.2: Automated determination of Hegman fineness by Chinese supplier Source: PERFECO Consulting Gysau

data collection, and reporting, also called hyper automation, plus collaborative robots (cobots) enhance human efforts for efficiency and precision.
- **AI-driven design of experiments** and simulations for faster innovation in combination with machine learning algorithms for predictive maintenance and data analysis will boost the innovation speed.
- The focus on **data-driven research** is based on the use of big data analytics for pattern recognition and hypothesis generation as well as digital twins for virtual simulations before practical experiments.
- **Remote operations** such as monitoring and control of equipment and experiments will increase reflecting new working models. Cloud-connected systems enabling access to data from anywhere and virtual reality and augmented reality interfaces for immersive collaboration will significantly increase.
- **Sustainability and energy efficiency** is for many industries a key innovation driver. This also demands renewable energy-powered lab operations. AI-optimized resource allocation leads to minimized waste.
- The **enhanced user experience** becomes more personalized. Modular systems will allow customization based on research requirements and scalable solutions for

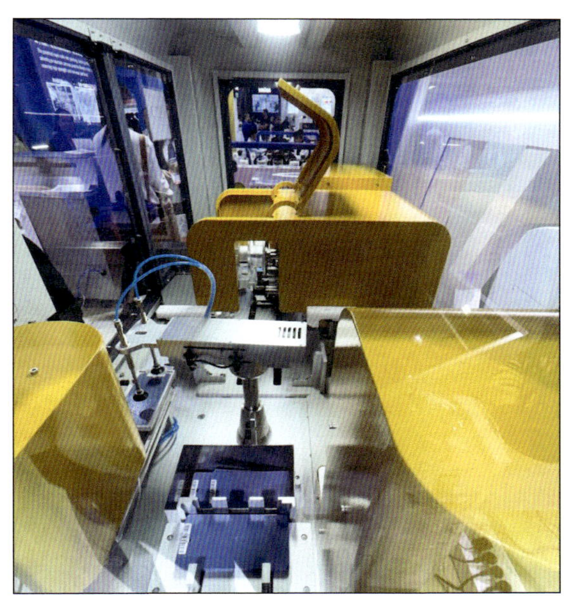

Figure 10.3: Fully automatic adhesion station from China
Source: PERFECO Consulting Gysau

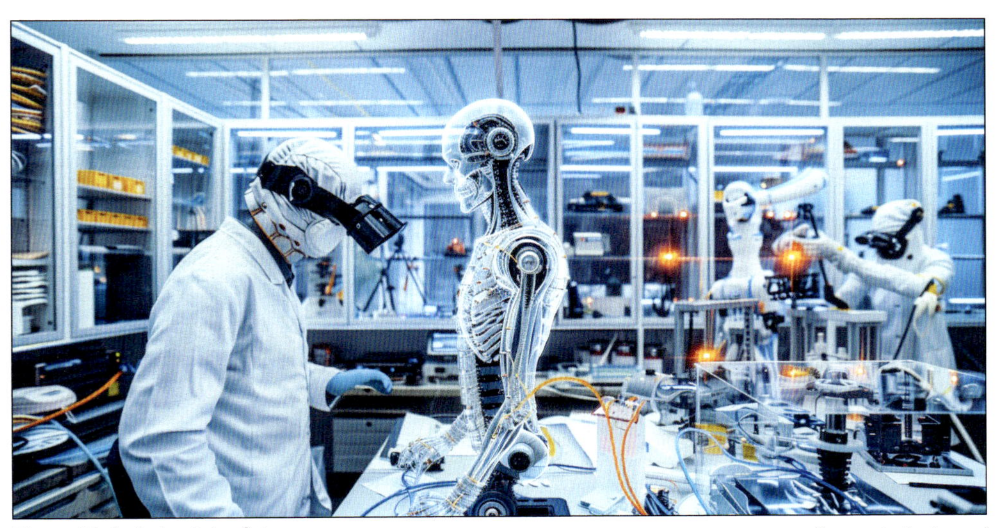

Figure 10.4: Lab of the future *Source: stockcake.com"*

labs of different sizes and industries. User interfaces will be intuitive for easy operation of advanced systems. AR/VR for training, troubleshooting, and interactive experiment guidance will become standard supporting voice-controlled systems for hands-free operation.

Smart labs represent the future of laboratory science, offering unprecedented capabilities through the integration of advanced technologies. By improving efficiency, precision, and sustainability, they empower researchers to achieve breakthroughs faster and more effectively. As the adoption of smart lab technologies grows, these innovations will become a cornerstone of R&D across industries.

10.4 Literature

[1] GYSAU, D.; "What's more important to enable efficient cyber-physical systems – digital twins or the degree of automation in R&D labs"; European Coatings Conference, 28th March 2023, Nuremberg

[2] KRIEGBAUM, E.; "We see a need to catch up"; www.european-coatings.com/news/markets-companies/interview-we-see-a-need-to-catch-up/, European Coatings, 13th Dec. 2022, Vincentz Network, Nuremberg

Author

Detlef Gysau, PERFECO Consulting Gysau, has been engaged in the development of hydro-fillers for the automotive industry at Akzo Coatings in Stuttgart, after his apprenticeship of a paint laboratory assistant (1985 - 88). During his studies he joined the industrial research centre of Rohm and Haas, Philadelphia and the R+D lab for photo initiators at Ciba-Geigy, Basel. 1996 he acquired his engineer degree in the fields of paints, lacquers and plastics (M.Eng.) at the University of Applied Science in Stuttgart. For over 14 years, he headed the Applied Technology Services for Paints, Coatings & Adhesives (ATS-PCA) division at Omya International, with global responsibility for development and technical services. During this period, he wrote his first book "Füllstoffe" (2005), followed by the English version "Fillers for Paints" in 2006, both published at Vincentz Network. In 2010 Detlef Gysau finished his Executive MBA in General Management at the University of St. Gallen, Switzerland and changed to Omya's Group Function Sales & Marketing. Here he held positions as Marketing Strategy Manager, Head of Product Management and Head of Innovation & Technical Marketing in the Segment Construction. In summer 2020 he followed as call by Holcim Technology as Head of Product Development Construction in the department Industrial Minerals in Holderbank, Switzerland. With a background in laboratory automation at Omya, Detlef Gysau joined Swiss-based Chemspeed Technologies in early 2022 as a member of the board and senior vice president for the Consumer Goods & Performance Materials markets. At the beginning of 2025 he decided to pursue his own consulting business in the field of raw materials, in particular mineral fillers, applications of paints, coatings, adhesives and building materials and last but not least lab automation in R&D and QC.

Index

Symbols

A

B

C

D

I

K

L

M

N

O

P

Index of the company names

Index of products